CRC SERIES IN CHROMATOGRAPHY

Gunter Zweig and Joseph Sherma
Editors-in-Chief

GENERAL DATA AND PRINCIPLES
Editors
Gunter Zweig, Ph.D.
U.S. Environmental Protection Agency
Washington, D.C.
Joseph Sherma, Ph.D.
Lafayette College
Easton, Pennsylvania

CARBOHYDRATES
Editor
Shirley C. Churms, Ph.D.
Research Associate
C.S.I.R. Carbohydrate Chemistry Research Unit
Department of Organic Chemistry
University of Cape Town, South Africa

DRUGS
Editor
Ram Gupta, Ph.D.
Department of Laboratory Medicine
St. Joseph's Hospital
Hamilton, Ontario, Canada

TERPENOIDS
Editor
Carmine J. Coscia, Ph.D.
Professor of Biochemistry
St. Louis University School of Medicine
St. Louis, Missouri

STEROIDS
Editor
Joseph C. Touchstone, B.S., M.S., Ph.D.
Professor
School of Medicine
University of Pennsylvania
Philadelphia, Pennsylvania

PESTICIDES AND RELATED ORGANIC CHEMICALS
Editors
Joseph Sherma, Ph.D.
Joanne M. Follweiler, Ph.D.
Lafayette College
Easton, Pennsylvania

LIPIDS
Editor
H. K. Mangold, Dr. rer. nat.
Executive Director and Professor
Federal Center for Lipid Research
Munster, Federal Republic of Germany

HYDROCARBONS
Editors
Walter L. Zielinski, Jr., Ph.D.
Air Program Manager
National Bureau of Standards
Washington, D.C.

INORGANICS
Editor
M. Qureshi, Ph.D.
Professor, Chemistry Section
Zakir Husain College of Engineering and Technology
Aligarh Muslim University
Aligarh, India

PHENOLS AND ORGANIC ACIDS
Editor
Toshihiko Hanai, Ph.D.
University of Montreal
Quebec, Canada

AMINO ACIDS AND AMINES
Editor
Dr. S. Blackburn
Leeds, England

POLYMERS
Editor
Charles G. Smith
Dow Chemical, USA
Midland, Michigan

PLANT PIGMENTS
Editor
Dr. Hans-Peter Köst
Botanisches Institüt der Universität München
München, Federal Republic of Germany

CRC Handbook of Chromatography

Terpenoids

Volume I

Carmine J. Coscia, Ph.D.
Professor of Biochemistry
St. Louis University School of Medicine
St. Louis, Missouri

Editors-in-Chief

Gunter Zweig, Ph.D.
School of Public Health
University of California
Berkeley, California

Joseph Sherma, Ph.D.
Professor of Chemistry
Chemistry Department
Lafayette College
Easton, Pennsylvania

CRC Press, Inc.
Boca Raton, Florida

Library of Congress Cataloging in Publication Data

Coscia, Carmine J.
Terpenoids.

(CRC handbook of chromatography)
Bibliography: p.
Includes index.
1. Terpenes--Analysis. 2. Chromatographic analysis. I. Zweig, Gunter. II. Sherma, Joseph.
III. Series.
QD416.C8 1984 v.1 547.7'1 83-14333
ISBN 0-8493-3004-1 (v. 1)

Direct all inquiries to CRC Press, Inc., 2000 Corporate Blvd., N.W., Boca Raton, Florida, 33431.

International Standard Book Number 0-8493-3004-1

Library of Congress Card Number 83-14333
Printed in the United States

EDITORS' PREFACE

The present volume in the series of Handbooks of Chromatography covers the broad field of terpenoids and has been admirably put together by Professor Carmine J. Coscia. The initial volumes of this series first appeared in 1972 as Volumes I and II, but the literature of specific topics that was covered in these first volumes has expanded to such a degree, that we felt it was more appropriate to issue separate volumes on specific topics each edited or authored by renowned experts in the field. Special volumes of this series, preceding Professor Coscia's treatment of terpenoids, include topics like drugs by Gupta, carbohydrates by Churms, polymers by Smith and co-workers, phenols and organic acids by Hanai, amines and amino acids by Blackburn, lipids and fatty acids by Mangold, and pesticides by Follweiler and Sherma. Other Handbooks of Chromatography are in preparation and will be published during the next several years, covering topics like peptides, inorganic compounds, plant pigments, steroids, and hydrocarbons.

As we have stated in the preface of previous volumes, we invite our devoted readers to comment on the contents of this volume and send their communications directly to the volume editor, Professor Coscia. We would also welcome suggestions from our readers about possible future volumes of this series and recommendations for possible editors. Professor Coscia's preface has very ably covered the scope of this Handbook, and any attempt by us to do the same would be superfluous.

Gunter Zweig, Ph.D.
Joseph Sherma, Ph.D.
Series Editors

THE EDITORS-IN-CHIEF

Gunter Zweig, Ph.D., received his undergraduate and graduate training at the University of Maryland, College Park, where he was awarded the Ph.D. in biochemistry in 1952. Two years following his graduation, Dr. Zweig was affiliated with the late R. J. Block, pioneer in paper chromatography of amino acids. Zweig, Block, and Le Strange wrote one of the first books on paper chromatography which was published in 1952 by Academic Press and went into three editions, the last one authored by Gunter Zweig and Dr. Joe Sherma, the co-Editor-in-Chief of this series. *Paper Chromatography* (1952) was also translated into Russian.

From 1953 till 1957, Dr. Zweig was research biochemist at the C. F. Kettering Foundation, Antioch College, Yellow Springs, Ohio, where he pursued research on the path of carbon and sulfur in plants, using the then newly developed techniques of autoradiography and paper chromatography. From 1957 till 1965, Dr. Zweig served as lecturer and chemist, University of California, Davis and worked on analytical methods for pesticide residues, mainly by chromatographic techniques. In 1965, Dr. Zweig became Director of Life Sciences, Syracuse University Research Corporation, New York (research on environmental pollution), and in 1973 he became Chief, Environmental Fate Branch, Environmental Protection Agency (EPA) in Washington, D.C.

During his government career, Dr. Zweig continued his scientific writing and editing. Among his works are (many in collaboration with Dr. Sherma) the now 11-volume series on *Analytical Methods for Pesticides and Plant Growth Regulators* (published by Academic Press); the pesticide book series for CRC Press; co-editor of *Journal of Toxicology and Environmental Health*; co-author of basic review on paper and thin-layer chromatography for *Analytical Chemistry* from 1968 to 1980; co-author of applied chromatography review on pesticide analysis for *Analytical Chemistry*, beginning in 1981.

Among the scientific honors awarded to Dr. Zweig during his distinguished career are the Wiley Award in 1977, Rothschild Fellowship to the Weizmann Institute in 1963/64; the Bronze Medal by the EPA in 1980.

Dr. Zweig has authored or co-authored over 75 scientific papers on diverse subjects in chromatography and biochemistry, besides being the holder of three U.S. patents.

At the present time (1980/84), Dr. Zweig is Visiting Research Chemist in the School of Public Health, University of California, Richmond, where he is doing research on farmworker safety as related to pesticide exposure.

Joseph Sherma, Ph.D., received a B.S. in Chemistry from Upsala College, East Orange, N.J., in 1955 and a Ph.D. in Analytical Chemistry from Rutgers University in 1958. His thesis research in ion exchange chromatography was under the direction of the late William Rieman, III. Dr. Sherma joined the faculty of Lafayette College in September 1958, and is presently Charles A. Dana Professor of Chemistry in charge of two courses in analytical chemistry. At. Lafayette he has continued research in chromatography and had additionally worked a total of 12 summers in the field with Harold Strain at the Argonne National Laboratory, James Fritz at Iowa State University, Gunter Zweig at Syracuse University Research Corporation, Joseph Touchstone at the Hospital of the University of Pennsylvania, Brian Bidlingmeyer at Waters Associates, and Thomas Beesley at Whatman, Inc., Clifton, N.J.

Dr. Sherma and Dr. Zweig (who is now with the University of California-Richmond) co-authored or co-edited the original Volumes I and II of the *CRC Handbook of Chromatography*, a book on paper chromatography, seven volumes of the series *Analytical Methods for Pesticides and Plant Growth Regulators*, and the Handbooks of Chromatography of drugs, carbohydrates, polymers, and phenols and organic acids. Other books in the pesticide series and further volumes of the *CRC Handbook of Chromatography* are being edited with Dr. Zweig, and Dr. Sherma has co-authored the handbook on pesticide chromatography. A book on quantitative TLC was edited jointly with Dr. Touchstone, and a general book on TLC was co-authored with Dr. B. Fried. Dr. Sherma has been co-author of eight biennial reviews of column liquid and thin layer chromatography (1968—1982) and the 1981 and 1983 reviews of pesticide analysis for the journal *Analytical Chemistry*.

Dr. Sherma has written major invited chapters and review papers on chromatography and pesticides in *Chromatographic Reviews* (analysis of fungicides), *Advances in Chromatography* (analysis of nonpesticide pollutants), Heftmann's *Chromatography* (chromatography of pesticides), Race's *Laboratory Medicine* (chromatography in clinical analysis), *Food Analysis: Principles and Techniques* (TLC for food analysis), *Treatise on Analytical Chemistry* (paper and thin layer chromatography), *CRC Critical Reviews in Analytical Chemistry* (pesticide residue analysis), *Comprehensive Biochemistry* (flat bed techniques), *Inorganic Chromatographic Analysis* (thin layer chromatography), and *Journal of Liquid Chromatography* (advances in quantitative pesticide TLC). He is editor for residues and elements for the JAOAC.

Dr. Sherma spent 6 months in 1972 on sabbatical leave at the EPA Perrine Primate Laboratory, Perrine, Fla., with Dr. T. M. Shafik, and two additional summers (1975, 1976) at the USDA in Beltsville, Md., with Melvin Getz doing research on pesticide residue analysis methods development. He spent 3 months in 1979 on sabbatical leave with Dr. Touchstone developing clinical analytical methods. A total of more than 230 papers, books, book chapters, and oral presentations concerned with column, paper, and thin layer chromatography of metal ions, plant pigments, and other organic and biological compounds; the chromatographic analysis of pesticides; and the history of chromatography have been authored by Dr. Sherma, many in collaboration with various co-workers and students. His major research area at Lafayette is currently quantitative TLC (densitometry), applied mainly to clinical analysis and pesticide residue and food additive determinations.

Dr. Sherma has written an analytical quality control manual for pesticide analysis under contract with the U.S. EPA and has revised this and the EPA Pesticide Analytical Methods Manual under a 4-year contract jointly with Dr. M. Beroza of the AOAC. Dr. Sherma has also written an instrumental analysis quality assurance manual and other analytical reports for the U.S. Consumer Product Safety Commission, and is currently preparing two manuals on the analysis of food additives for the U.S. FDA, both of these projects also in collaboration with Dr. Beroza of the AOAC.

Dr. Sherma taught the first prototype short course on pesticide analysis with Henry Enos of the EPA for the Center for Professional Advancement. He was editor of the Kontes TLC quarterly newsletter for 6 years and also has taught short courses on TLC for Kontes and the Center for Professional Advancement. He is a consultant for numerous industrial companies and federal agencies on chemical analysis and chromatography and regularly referees papers for analytical journals and research proposals for government agencies. At Lafayette, Dr. Sherma, in addition to analytical chemistry, teaches general chemistry and a course in thin layer chromatography.

Dr. Sherma has received two awards for superior teaching at Lafayette College and the 1979 Distinguished Alumnus Award from Upsala College for outstanding achievements as an educator, researcher, author, and editor. He is a member of the ACS, Sigma Xi, Phi Lambda Upsilon, SAS, and AIC.

PREFACE

In keeping with previous volumes of this series, the purpose of this handbook is to provide data on the resolution of terpenoids by modern chromatographic techniques. This is accomplished in three sections that deal with: (1) tabulated chromatographic mobilities, (2) methods of detection of terpenoids, and (3) techniques of sample preparation and prefractionation of tissue and biological fluid extracts. Although most classes of isoprenoids including terpenoid quinones and polyisoprenoid alcohols are treated, due to space considerations, steroids and vitamin D will not be covered. (Steroid chromatography is the subject of a separate volume in this series.)

In the first section, the chromatographic behavior of terpenes when subjected to gas, liquid, and thin layer chromatography are presented. For each table, citations to the original literature are included. Since most terpenes are more readily resolved in these systems, there have been only a few applications of paper chromatography. In some cases in which the latter method could have been used for very polar terpenoids, it has been replaced by the more rapid procedure of thin layer chromatography on cellulose. Each subsection is comprised of tables in which retention times are provided for a specific class of terpenes. In most instances, the tables are restricted to terpenes with the same number of isoprenoid units. They are presented in succession beginning with monoterpenes and ending with polyisoprenoid alcohols, carotenoids, and isoprenoid quinones in that order. A few tables contain a cross-section of representative terpenes with different numbers of isoprenoid groups.

In the second major section, reagents used to detect terpenes are described. Most of these reagents have been developed for thin layer chromatography. Some are very specific for certain terpenes, but many will detect other lipids as well. Color specificity is also recorded.

Finally, the last section deals with various techniques adopted to prefractionate tissues or biological fluids, before the chromatographic methods found in the first section can be applied. This is of particular importance in the clinical sciences and chemistry of natural products where one wishes to resolve terpenoids from complex mixtures. As in the first section, the methods are listed according to the group(s) of terpenes for which they are most useful, beginning with monoterpenes. As the terpenoids increase their number of isoprenoid units, they change from volatile molecules to high-melting solids so that the methods of tissue work-up can vary considerably.

This volume covers the most recent literature on the chromatography of terpenoids beginning with 1970. Only a few important earlier publications are cited. It is designed to be of use for research chemists and biochemists interested in the isolation of terpenoids from various sources and clinical analysis of the fat-soluble vitamins A, E, and K. It is intended not only for the detection of previously identified terpenoids, but the application of the techniques should be of value in the isolation of new compounds as well. Finally we hope that the methods outlined will assist plant enzymologists in their investigations of biological transformations of terpenoids by providing simple effective means of resolving these lipids.

I would like to express my gratitude to my wife, Jo, my son, Tom, and my secretary, Catherine Mack, for their assistance in preparing this handbook.

C. J. Coscia

THE AUTHOR

Chromatography has been a mainstay in the research of Dr. Carmine J. Coscia throughout his scientific career. His contributions have been on the forefront of new developments and applications of chromatographic methodology. In 1953 he was awarded a scholarship to Manhattan College in New York for a research project on the paper chromatography of inorganic ions. He obtained a B.S. degree in Chemistry there, graduating magna cum laude in 1957. He then attended Fordham University where he completed his Ph.D. dissertation on lignification in higher plants with Dr. F. F. Nord. In this period Dr. Coscia introduced the use of gas-liquid chromatography in the resolution of lignin hydrogenation products. The next two years (1962-1964) were spent as a postdoctoral fellow in the laboratory of Dr. D. Arigoni at the Eidgenossische Technische Hochschule in Zurich, Switzerland. While studying the biosynthesis of terpenoids, he exploited both thin layer as well as direct and reverse-phase liquid chromatography in the isolation of new intermediates. Upon returning to the United States, he joined the laboratory of Dr. Ronald Bentley at the University of Pittsburgh. In their collaboration on the biosynthesis of vitamin K and ubiquinone in microorganisms, both thin layer and liquid chromatography were adopted to identify these natural products as well as their aromatic precursors.

In 1965 he was appointed Assistant Professor in the E. A. Doisy Department of Biochemistry at St. Louis University School of Medicine. He was promoted to Associate Professor in 1970 and Full Professor in 1973, his present position. He received a Career Development Award from the National Institutes of Health from 1968 to 1972. In this period he has continued to advance new applications of chromatographic methods in the study of both plant and animal isoprenoid and aromatic amino acid metabolism. Among his more than 50 published papers are included contributions on paired-ion reverse-phase high pressure liquid chromatography and computerized gas liquid chromatography-mass spectrometry of aromatic amines.

Dedicated to

Jo

ADVISORY BOARD

TABLE OF CONTENTS

INTRODUCTION

Terpenoids are a ubiquitous, structurally diverse group of lipids that exist in a variety of polymeric forms of the unsaturated, branch-chain pentane, isoprene. A nomenclature for terpenes has evolved in which the dimer of isoprene was established as the fundamental unit, reflecting the preponderance of this class in higher plants. As a result, terpenes are categorized as hemiprenes (C_5), monoterpenes (C_{10}), sesquiterpenes (C_{15}), diterpenes (C_{20}), triterpenes (C_{30}), etc.

Terpenoids are found in cyclic and linear forms, with the latter containing as many as 22 isoprenyl residues. The cyclic compounds exist in a vast array of stereoisomers. In addition, linear polyprenoids exhibit geometrical isomerism of the double bond present in their isoprene units. In considering their distribution in nature, it is possible to classify terpenoids into two groups. The first includes those compounds that occur more or less consistently in the plant kingdom as primary metabolites. The second encompasses the secondary metabolites, i.e., terpenoids that are species-specific.

Among the first class are the carotenoids, triterpenoids, terpenoid quinones, polyisoprenoids, and several growth regulators. With some minor variations the carotenoids common to most plants possess similar structural features, 40 carbons and a definitive function in photosynthetic electron transport.[1] Even more ubiquitous are the terpenoid quinones, such as ubiquinones, menaquinones, plastoquinones, and quinones related to vitamin E.[2] They function as electron carriers in bacteria, plants, and animals. Similarly, triterpenoids that are membrane components in plants appear to be the counterpart of cholesterol in animal cells.[3] Whether they perform identical roles remains to be established. Polyisoprenoid alcohols are the most universally distributed isoprenoids, occurring across the gamut of living organisms from the simplest bacterial forms to mammals.[4] They serve as carriers of carbohydrates in the biosynthesis of glycoproteins and glycolipids in cell membranes. It is thought that the lipophilic polyisoprenoids covalently bind to hydrophilic sugar phosphates to facilitate their transfer in the lipid milieu, wherein complex oligosaccharide assembly occurs.

Terpenoids often play important hormonal roles in eukaryotes. Although the diterpene gibberellins were originally discovered in the fungus, *Fusarium moniliforme*, considerable evidence exists to suggest a broad distribution in plants.[5] Gibberellins are plant hormones that increase membrane proliferation and protein synthesis, possibly affecting the latter at the transcriptional level. Abscissic acids, which can counteract the effect of gibberellins in plants,[5] are sesquiterpene hormones that are structurally similar to juvenile hormones in insects.[6] Both sesquiterpene hormones are widely distributed in their respective kingdoms. Cytokinins, which contain a hemiprene unit attached to the 6-amino group of the purine, adenine, have been detected in bacteria, yeast, plants, and animals.[7,8] They are incorporated into transfer-RNA. In general N^6-(Δ^2 isopentenyl) adenosine and its analogues act to promote nucleic acid synthesis and hence growth.

The other major classification of terpenoids encompass the secondary metabolites whose distribution are highly species-dependent and whose function is undefined, but likely to be allochemical in nature.[9] Many terpenoids of this group have economic importance, and the cultivation of plants bearing these compounds is widespread. For example, the essential oils consist of a complicated mixture of mono-, sesqui-, and diterpenes that have been used for centuries as medicinals, condiments, fragrances, and solvents. Some terpenoids have proven to be useful experimental agents. The phorbols are a group of tetracyclic diterpenes prevalent in croton species that have potent synergistic effects on carcinogens in tumor promotion.[10] Another diterpene, forskolin, is a unique activator of brain and cardiac adenylate cyclases.[11] Furthermore, terpenoids are frequently found in nature as one moiety of other natural products. Noteworthy are the monoterpenoid indole alkaloids[12] and the hemiprene-containing ergot alkaloids,[13] which have seen extensive medicinal application.

A broad distribution of terpenoid secondary metabolites exists in plants, and their functional significance for the most part has remained an enigma. Clearly, allochemical roles are most attractive possibilities for volatile monoterpenes, and evidence for this line of reasoning has been gained in the study of insect-plant interrelationships.[9] In addition, certain insects store and secrete monoterpenes, as a defense mechanism against other insects.[14] Needless to say, terpenes have been implicated in animal-plant, microbial-plant, and plant-plant interactions as well.[9] In addition to the deterrent action of volatile terpenes,[15,16] there may be roles for nonvolatile isoprenoids. For example, previously mentioned diterpenoids, the phorbols which are skin irritants and terpenoid alkaloids which are toxic, may serve as defensive chemical messengers in the communication between plants and animals. Conversely, it has been speculated that the heterogeneity of terpenoid populations in plants reflects evolutionary divergency.[3] Consistent with this line of reasoning would be a putative multiplicity of functions such as hormonal, etc., that are unique to each species.

Although their function remains to be established, the molecular diversity of terpenes has provided many advantages as well as challenges. For chemists with an interest in structure elucidation and stereochemical analyses, terpenes were one of the first groups of natural products to be investigated.[17] The brilliant insight of Ruzicka[18] in postulating the isoprene biogenetic rule, which recognized the basic regularity of the isoprenoid molecule, represented a major breakthrough of that period. With skills developed in the structure elucidation of terpenes, natural product chemists were able to turn to other secondary metabolites with excellent models to draw upon. They were also confronted with molecules that challenge the ability of the chemists to synthesize complex structures.[17]

The structural diversity of terpenes, often brought about by subtle molecular changes, makes them excellent subjects for spectroscopic analyses. Indeed, different plants often provide enantiomeric forms of terpenes with several centers of asymmetry. Spectral comparisons of enantiomers can afford insight into the diagnostic potential of the particular method. Excellent exploitation of terpenes has recently been accomplished in the development of ^{13}C-magnetic resonance spectroscopy[19] and mass spectrometry.[20]

The study of the biosynthesis of isoprenoids was pioneered by Bloch, Lynen, Popjak, and Cornforth, who focused on the formation and metabolism of cholesterol in mammals.[21] Their classic investigations represented the prototype for subsequent research not only on the biosynthesis of terpenes in plants and microorganisms but for other natural products as well. Now that the general features of this biosynthetic pathway have been resolved, details of the individual steps are being scrutinized in enzymatic studies.[21] Interestingly, the biogenesis of terpenes has run a full circle. In most investigations the structure of a terpenoid has served as a guide for the elucidation of its biosynthesis.[21] Recently, a structure elucidation of unknown terpenes was abetted by tracer studies. The incorporation of radioactive label from precursors such as acetate and mevalonate into terpenes of unknown structures followed by degradation to known products (e.g., Kuhn-Roth oxidation) was used to gain insight into their structure.[22]

Critical to the successful use of terpenoids for diagnosis of molecular properties, as well as for biochemical studies, has been the ability to isolate pure compounds. In some cases a terpene metabolite will constitute a major component of a specific tissue. Astonishingly, simple isolation procedures involving solvent extraction or steam distillation followed by fractional crystallization have yielded enormous amounts of a crystalline homogeneous terpene from an abundant plant tissue. The isolation of camphor from the bark of *Cinnamomum camphor* is a well known example. Nevertheless, this is exceptional and more often than not, chemists have been confronted with the task of resolving highly complicated mixtures exemplified by essential oils. This has stimulated the development of more sophisticated techniques than distillation or fractional crystallization. In fact, terpenes have provided a potent stimulus to the beginning of modern chromatography. One of the first mixtures to be subjected to column chromatography were carotenoids. The chromatography of terpenoids has been an integral part of modern chemistry and remains critical to a myriad of commercial processes such as the production of terpenoid medicinals, flavors, and fragrances.

The recent surge of interest in the bioregulatory functions of terpenoids such as vitamins A, E, and K in human physiology has extended the application of terpenoid chromatography into the realm of clinical chemistry. The A vitamins are not only required in the visual process, but a growing body of evidence suggests vitamin A plays a role in growth, reproduction and epithelial differentiation.[23] Analogues of vitamin A have been successfully used in the treatment of skin diseases as well as different forms of epithelial cancer.[24,25] The nutritional benefits of vitamin E have not shed a great deal of light on its function.[26] Vitamin E promotes erythrocyte stability, but recent research including the discovery of a possible α-tocopherol binding protein in rat liver cytoplasm belie the hypothesis that it functions as a simple antioxidant alone.[26] Vitamin K has now been firmly implicated in the posttranslational modification of specific proteins such as prothrombin. It probably functions as a redox cofactor in the γ-carboxylation of glutamate residues by either simple electron transport or by a more complicated mechanisms.[28] The resulting malonic acid residues impart to the protein a high affinity for calcium.

Critical to interpreting the function of these vitamins in mammals is the additional knowledge of their distribution and metabolism, the study of which will depend on modern chromatographic techniques featuring high resolution and sensitivity. A major part of this book is devoted to describing these techniques.

REFERENCES

1. **Goodwin, T. W.,** *The Biochemistry of Carotenoids,* Vol. 1, 2nd ed., Chapman and Hall, London, 1981.
2. **Threlfall, D. R.,** *Biochim. Biophys. Acta,* 280, 472, 1972.
3. **Ourisson, G., Rohmer, M., and Anton, R.,** *Recent Adv. Phytochem.,* 13, 131, 1979.
4. **Lennarz, W. J.,** *Science,* 108, 986, 1975.
5. **Varner, J. E. and Ho, D. T.,** in *Plant Biochemistry,* 3rd ed., Bonner, J. and Varner, J. E., Eds., Academic Press, New York, 1976, 714.
6. **Gilbert, L. I.,** *The Juvenile Hormones,* Plenum Press, New York, 1976.
7. **Horgan, R.,** *Proc. R. Soc. London Ser. B,* 284, 439, 1978.
8. **Laloue, M.,** *Proc. R. Soc. London Ser. B,* 284, 449, 1978.
9. **Whittaker, R. H. and Feeny, P. P.,** *Science,* 71, 757, 1971.
10. **Hecker, E. and Schmidt, R.,** *Prog. Chem. Org. Nat. Prod.,* 31, 377, 1974.
11. **Seamon, K. B., Padgett, W., and Daly, J. W.,** *Proc. Natl. Acad. Sci. U.S.A.,* 78, 3363, 1981.
12. **Cordell, G. A.,** *Lloydia,* 37, 219, 1974.
13. **Floss, H. G.,** *Tetrahedron,* 32, 873, 1976.
14. **Cavill, G. W. and Clark, D. V.,** in *Naturally Occurring Insecticides: Juvenile Hormones,* Jacobson, M. and Crosby, D. G., Eds., Marcel Dekker, New York, 1971.
15. **Bryant, J. P.,** *Science,* 213, 889, 1981.
16. **Farentinos, R. C., Capretta, P. J., Kepner, R. E., and Littlefield, V. M.,** *Science,* 213, 1273, 1981.
17. **Newman, A. A.,** *Chemistry of Terpenes and Terpenoids,* Academic Press, New York, 1972.
18. **Ruzicka, L.,** *Experientia,* 9, 357, 1953.
19. **Wehrli, F. W. and Wirthlin, T.,** *Interpretation of Carbon-13 NMR Spectra,* Heyden and Sons, London, 1978.
20. **Enzell, C. R. and Wahlberg, I.,** in *Biochemical Applications of Mass Spectrometry,* First Supplementary Volume, Waller, G. R. and Dermer, O. C., Eds., John Wiley & Sons, New York, 1980, 311.
21. **Porter, J. W. and Spurgeon, S. L., Eds.,** *Biosynthesis of Isoprenoid Compounds,* Vol. 1, John Wiley & Sons, New York, 1981.
22. **Arigoni, D.,** unpublished observations.
23. **Sani, B. P.,** *Biochem. Biophys. Res. Commun.,* 75, 7, 1977.

24. **Sporn, M. B., Dunlop, N. M., Newton, D. L., and Smith, J. M.,** *Fed. Proc.*, 35, 1332, 1976.
25. **Mayer, H., Bollag, W., Hänni, R., and Rüegg, R.,** *Experientia,* 34, 1105, 1978.
26. **Bieri, J. G. and Farrell, P. M.,** *Vitam. Horm. (N.Y.),* 34, 31, 1976.
27. **Catignani, G. L. and Bieri, J. G.,** *Biochim. Biophys. Acta,* 497, 349, 1977.
28. **Olson, R. E. and Suttie, J. W.,** *Vitam. Horm. (N.Y.),* 35, 59, 1977.

Section I
Tables of Chromatographic Data

Table GC1
MONOTERPENE HYDROCARBONS

Column packing	P1	P2	P3
Temperature (°C)	T1	T2	T3
Gas; Flow rate (mℓ/min)	He; 3	He; 20	He; 65
Column			
Length (ft)	50	10	10
Diameter	0.02 in. I.D.	2 mm I.D.	0.25 in. O.D.
Form	Capillary	na	na
Material	na	Glass	SS
Detector	Mass spec.	Mass spec.	Mass spec.
Reference	1	2	3

Compound	$r_{limonene} \times 10^3$	$r_{n\text{-heptyl acetate}}$ [a]	t_r (min)
Cyclofenchene	469	—	—
Santene	539	0.221 ± 0.039	—
Tricyclene	597	0.268 ± 0.039	—
α-Thujene	614	—	—
α-Pinene	627	0.298 ± 0.039	21.97
β-Fenchene	659	—	—
α-Fenchene	702	—	—
Camphene	722	0.364 ± 0.039	—
β-Pinene	822	0.446 ± 0.039	34.50
Myrcene	857	0.525 ± 0.039	—
Δ^3-Carene	899	0.548 ± 0.039	—
α-Phellandrene	922	0.549 ± 0.039	—
α-Terpinene	952	0.592 ± 0.039	40.80
Limonene	1000	0.613 ± 0.039	42.77
Sabinene	1028	—	—
β-Phellandrene	1028	0.637 ± 0.039	50.63
β-Ocimene-x	1028	—	—
β-Ocimene-y	1058	—	—
Cymene	1078	0.736 ± 0.039	50.63
γ-Terpinene	1103	—	47.07
Terpinolene	1188	0.815 ± 0.039	—
allo-Ocimene(*trans-cis*?)	1313	—	—
allo-Ocimene(*trans-trans*?)	1366	—	—
1,8-Cineole	—	0.658 ± 0.039	—
Cis-Ocimene	—	0.686 ± 0.015	—
Trans-Ocimene	—	0.718 ± 0.038	—

[a] RRT and 5-standard deviation windows.

Column packing
- P1 = Polyphenyl ether
- P2 = 1% Polyethylene glycol 20M + 1% OV-17 on Chromosorb G (80-100 mesh)
- P3 = 20% LAC 446 on 60-80 mesh Chromosorb W

Temperature
- T1 = Program, 8 min at 60°, 60 to 150° at 2°/min
- T2 = Program, 60° to 190° at 4°/min
- T3 = Program, 50° for 50 min then increase at 1°/min for 105 min, then 155°C for 25 min

REFERENCES

1. **Tyson, B. J.,** *J. Chromatogr.*, 111, 419, 1975.
2. **Adams, R. P., Granot, M., Hogge, L. R., and von Rudloff, E.,** *J. Chromatogr. Sci.*, 17, 75, 1979.
3. **Kumamoto, J., Waines, J. G., Hollenberg, J. L., and Scora, R. W.,** *J. Agric. Food Chem.*, 27, 203, 1979.

Table GC 2
MONOTERPENE HYDROCARBONS

Column packing	P1	P2	P3
Temperature (°C)	80	80	80
Gas; Flow rate (cm/sec)	H_2; 8.7	H_2; 8.7	H_2; 8.7
Column			
Length(mm)	5	5	5
Diameter (mm I.D.)	1.8	1.8	1.8
Form	Coiled	Coiled	Coiled
Material	Glass	Glass	Glass
Detector	na	na	na

Compound	**$V_R/(V_R)$Limonene × 10^3**		
Tricyclene	300	—	310
α-Pinene	350	400	360
α-Thujene	400	—	400
Camphene	425	450	430
β-Pinene	535	550	535
Sabinene	700	610	700
Δ^3-Carene	730	790	750
Myrcene	865	800	830
α-Phellandrene	890	860	860
α-Terpinene	975	965	990
β-Phellandrene	1100	1010	1070
1,8-Cineole	1195	1170	1180
1,4-Cineole	1205	1300	1290
γ-Terpinene	1250	1200	1270
Terpinolene	1300	1360	1320
p-Cymene	1550	—	—
cis-Ocimene	1600	1220	1430

Column packing P1 = 1.5% 1,2,3-Tris-(2-cyanoethoxy) propane (TCEP) on Carbopack C (100-120 mesh)

P2 = 1.0% *N,N,N',N'*-tetrakis-(2-hydroxyethyl) ethylene-diamine (THEED) on Carbopack C (100-120 mesh)

P3 = 0.9% TCEP + 0.5% THEED on Carbopack C (100-120 mesh)

REFERENCE

DiCorcia, A., Liberti, A., Sambucini, C., and Samperi, R., *J. Chromatogr.*, 152, 63, 1978.

Table GC 3
MONOTERPENES

Column packing	P1	P2
Temperature (°C)	32-130°, 6°/min	80-130°, 6°/min
Gas; Flow rate (mℓ/min)	He; 50	He; 50
Column		
Length (ft)	6	6
Diameter (mm, I.D.)	2	2
Form	na	na
Material	Glass	Glass
Detector	FI	FI

Compound	$r_{limonene}$	
α-Pinene	0.46	0.37
Camphene	0.51	0.49
β-Pinene	0.63	0.60
2-Methyl-2-heptene-6-one	0.62	
Myrcene	0.77	0.78
Δ^7-Carene	0.86	0.69
α-Terpinene	0.90	0.89
Limonene	1.00	1.00
β-Phellandrene	1.00	1.07
cis-Ocimene	1.14	1.19
trans-Ocimene	1.24	1.32
γ-Terpinene	1.28	1.29
Terpinolene	1.63	1.59
Linalool	1.84	

Column packing P1 = 3% OV-101 on Gas Chrom Q (100-120 mesh)
P2 = 20% Reoplex 400 on Chromosorb W (80-100) AW

REFERENCE

Hood, L. V. S., Dames, M. E., and Barry, G. T., *Nature (London)*, 242, 402, 1973.

Table GC 4
HYDROGENATED MONOTERPENES

Column packing		P1	P2
Temperature (°C)		65	100
Gas; Flow rate (mℓ/min)		H_2; 25	H_2; 25
Column			
Length (ft)		500	500
Diameter (in., I.D.)		0.03	0.03
Form		Capillary	Capillary
Material		SS	SS
Detector		FI	FI
Compound		**I**	
Acyclic			
Myrcene	2,6-dimethyloctane	938	922
Linalool	2,6-dimethyloctane	939	922
Citronellol	2,6-dimethylheptane	834	825
Citrenellol	2,6-dimethyloctane	938	923
Monocyclic			
Terpinolene	*trans-p*-menthane	981	1022
Terpinolene	*cis-p*-menthane	995	1045
Terpinolene	*p*-cymene	1018	1260
p-Cymene	*trans-p*-menthane	980	1022
p-Cymene	*cis-p*-menthane	994	1046
p-Cymene	*p*-cymene (unreacted)	1016	1260
α-Terpineol	*trans-p*-menthane	981	1022
α-Terpineol	*cis-p*-menthane	995	1045
α-Terpineol	*p*-cymene	1018	1261
1,8-Cineole	*trans-p*-menthane	981	1022
1,8-Cineole	*cis-p*-menthane	995	1045
1,8-Cineole	*p*-cymene	1018	1260
Bicyclic			
α-Pinene	*trans*-pinane	973	1049
α-Pinene	*cis*-pinane	983	1061
β-Pinene	*trans*-pinane	973	1049
β-Pinene	*cis*-pinane	983	1061
Camphene	*trans*-isocamphane	975	1056
Camphene	*cis*-isocamphane	980	1065
3-Carene	1,1,4-trimethylcycloheptane	972	1020
cis-Carane	1,1,4-trimethylcycloheptane	972	1021
cis-Carane	*trans-p*-menthane	981	1021
cis-Carane	*cis*-carane (unreacted)	986	1064
cis-Carane	*cis-p*-menthane	994	1045
cis-Carane	*m*-cymene	1011	1261
cis-Carane	*p*-cymene	1016	1261
Sabinene	1-*trans*-3-dimethyl-1-isopropylcyclopentane	945	976
α-Thujone	1-*trans*-2-dimethyl-*cis*-3-isopropylcyclopentane	933	953
α-Thujone	1-*trans*-3-dimethyl-1-isopropylcyclopentane	944	977
α-Thujone	1-*trans*-2-dimethyl-*trans*-3-isopropylcyclopentane	949	—
α-Thujone	1-*cis*-2-dimethyl-*trans*-3-isopropylcyclopentane	960	997
α-Thujone	1-*cis*-2-dimethyl-*cis*-3-isopropylcyclopentane	974	1015
Tricyclic			

Table GC 4 (continued)
HYDROGENATED MONOTERPENES

Compound		I	
Tricyclene	camphane	953	1021
Tricyclene	*trans*-isocamphane	975	1056
Tricyclene	*cis*-isocamphane	981	1066

Column packing P1 = SF 96(50)
P2 = Carbowax 20M

REFERENCE

Kepner, R. E. and Maarse, H., *J. Chromatogr.*, 66, 229, 1972.

Table GC 5
ACYCLIC MONOTERPENE ALCOHOLS

Column packing	P1	P2	
Temperature (°C)	150	150	
Gas; Flow rate (mℓ/min)	N_2; 20	N_2; 20	
Column			
Length (m)	2	2	
Diameter (mm, I.D.)	2	2	
Form	na	na	
Material	Glass	Glass	
Detector	FI	FI	
Compound	**I**	**I**	**ΔI**
Tetrahydrogeraniol	1145	1675	530
6,7-Dihydronerol	1151	1725	574
6,7-Dihydrogeraniol	1172	1759	587
Citronellol	1173	1765	592
α-Citronellol	1164	1760	596
Nerol	1175	1808	633
Geraniol	1189	1842	653
cis-Isogeraniol	1179	1812	633
trans-Isogeraniol	1184	1812	628
α-Isogeraniol	1169	1800	631
Myrcenol-8	1196	1919	727
Tetrahydrolavandulol	1123	1600	477
Lavandulol	1135	1707	572
Tetrahydromyrcenol	1067	1449	382
6(10)-Dihydromyrcenol	1036	1473	437
Myrcenol	1084	1631	547
cis-Ocimenol	1115	1660	545
trans-Ocimenol	1132	1685	553
Tetrahydrolinalool	1076	1431	355
6,7-Dihydrolinalool	1035	1449	414
1,2-Dihydrolinalool	1104	1537	433
Linalool	1064	1555	491

Column packing P1 = 20% Apiezon L on 60-80 mesh Embacel
P2 = 20% Carbowax 20M on 60-80 mesh Embacel

REFERENCE

Ter Heide, R., *J. Chromatogr.*, 129, 143, 1976.

Table GC 6
OXYGENATED MONOTERPENES

Column packing	P1	P2
Temperature (°C)	T1	T2
Gas; Flow rate (mℓ/min)	He; 20	N_2; 15
Column		
Length	10 ft	150 cm
Diameter	2 mm I.D.	0.04 cm
Form	na	na
Material	Glass	Glass
Detector	Mass spec.	FI
Reference	1	2

Compound	$r_{n\text{-heptyl acetate}}$[a]	t_r (sec)
Fenchone	1.034 ± 0.044	—
Thujone	1.084 ± 0.043	—
Isothujone	1.124 ± 0.011	—
Citronellal	1.210 ± 0.021	—
Fenchyl acetate	1.270 ± 0.069	—
Linalool	1.274 ± 0.026	—
Camphor	1.303 ± 0.079	—
Linalyl acetate	1.421 ± 0.029	—
Terpinen-4-ol	1.439 ± 0.099	—
Methyl thymol	1.480 ± 0.063	—
Bornyl acetate	1.528 ± 0.109	—
Pulegone	1.583 ± 0.040	270
Borneol	1.614 ± 0.053	—
α-Terpineol	1.626 ± 0.129	—
Verbenone	1.656 ± 0.129	—
Citronellyl acetate	1.695 ± 0.149	—
α-Terpinyl acetate	1.731 ± 0.054	—
Carvone	1.742 ± 0.057	270
Piperitone	1.755 ± 0.083	292
Citronellol	1.776 ± 0.169	—
Myrtenol	1.796 ± 0.199	—
Geranyl acetate	1.885 ± 0.199	—
Thymol	2.583 ± 0.151	—
Menthone	—	191
Dihydrocarvone	—	227
Carvenone	—	295
Piperitenone	—	415

[a] RRT and 5-standard deviation windows.

Column packing	P1 =	1% Polyethylene glycol 20M + 1% OV on Chromosorb G (80-100 mesh)
	P2 =	10% SE-30 on Diatomite C
Temperature	T1 =	60°C-190°C at 4°/min
	T2 =	151°C

REFERENCES

1. **Adams, R. P., Granat, M., Hogge, L. R., and von Rudloff, E.,** *J. Chromatogr. Sci.*, 17, 75, 1979.
2. **Rothbacher, H. and Suteu, F.,** *J. Chromatogr.*, 100, 236, 1974.

Table GC 7
CAMPHOR METABOLITES

Column packing	P1	P2
Temperature (°C)	160	225
Gas; Flow rate (mℓ/min)	N_2; na	He; 100
Column		
Length	na	6 ft.
Diameter	na	na
Form	na	na
Material	na	na
Detector	FI	FI
Reference	1	2
Compound	**t_r (min)**	
(±)-*cis*-2-*exo*,3-*exo*-Camphanediol	9.0	—
(±)-*cis*-2-*endo*,3-*endo*-Camphanediol	9.0	—
(±)-*trans*-2-*exo*,3-*endo*-Camphanediol	12.0	—
(±)-*trans*-2-*endo*,3-*exo*-Camphanediol	12.0	—
(±)-2-*endo*-Hydroxyepicamphor	3.5	—
(±)-3*endo*-Hydroxyepicamphor	3.5	—
(±)-5-*exo*-Hydroxycamphor	—	5.4
(±)-5-*endo*-Hydroxycamphor	—	6.6

Column packing P1 = 12% Carbowax 20M on Chromosorb P (100-120 mesh)
P2 = 20% Carbowax 20M on Chromosorb W (60-80 mesh)

REFERENCES

1. **Robertson, J. S. and Solomon, E.,** *Biochem. J.*, 121, 503, 1971.
2. **Leibman, K. C. and Ortiz, E.,** *Drug Metab. Dispos.*, 1, 543, 1973.

Table GC 8
MENTHOLS

Column packing	P1	P2
Temperature (°C)	T1	T1
Gas; Flow rate (mℓ/min)	N_2; 25	N_2; 25
Column		
Length (ft)	9	9
Diameter (in., O.D.)	2	2
Form	na	na
Material	Glass	Glass
Detector	EC	EC
Compound	**Methylene**	**Units**
Menthol		
p-Menthan-3-ol	13.41	12.61
3-Trimethylsilyloxy	12.52	12.66
Neomenthol		
p-Menthan-3-ol	13.32	12.48
3-Trimethylsilyloxy	12.25	12.49
Isomenthol		
p-Menthan-3-ol	13.41	12.61
3-Trimethylsilyloxy	12.41	12.46
Neoisomenthol		
p-Menthan-3-ol	13.51	12.72
Trimethylsilyloxy	12.65	12.67
Mentheglycol		
p-Menthan-3.8-diol (*trans*)	13.14	—
3-Trimethylsilyloxy	14.10	14.88
3-Trimethylsilyloxy	14.55	—
3,8-Di-(trimethylsilyloxy)	15.35	15.00
p-Cyclobutylboronate	14.95	15.93
3-Mono-(heptafluorobutyrate)	11.96	12.02
3,8-Di-(heptafluorobutyrate)	13.88	13.29
Neomenthoglycol		
p-Menthan-3,8-diol (*cis*)	12.92	—
3-Trimethylsilyloxy	13.79	14.49
3-Trimethylsilyloxy	14.34	14.88
3,8-Di-(trimethylsilyloxy)	15.12	14.81
p-Cyclobutylboronate	14.95	15.95
3-Mono-(heptafluorbutyrate)	11.86	11.92
3,8-Di-(heptafluorobutyrate)	13.45	12.79

Column packing P1 = 5% SE-30 on Gas Chrom P (120-140 mesh) AWS
P2 = 5% OV-17 on Gas-Chrom P (120-140 mesh) AWS
Temperature T1 = Program 1°/min from 70°

REFERENCE

Bournot, P., Maume, B. F., and Baron, C., *J. Chromatogr.*, 57, 55, 1971.

Table GC 9
LIMONENE AND METABOLITES

Column packing	P1	P2	P3
Temperature (°C)	120	110	200
Gas; Flow rate (mℓ/min)	N_2; 15	N_2; 18	N_2; 17.5
Column			
Length (m)	1.5	0.75	1.5
Diameter (mm, I.D.)	2.0	2.0	2.0
Form	na	na	na
Material	Glass	Glass	Glass
Detector	FI	FI	FI
Compound	**t_r (min)**		
d-Limonene	0.78	0.60	1.13
p-Mentha-1,8-dien-6-ol	1.34	1.10	2.25
p-Mentha-1,8-dien-10-ol	1.99	2.05	3.20
p-Mentha-1-ene-6,8,9-triol	2.12	2.36	3.70
p-Menth-l-ene-8,9-diol	4.01	4.79	6.05

Column packing P1 = 1.15% OV-1 on Shimalite W
P2 = 1.5% SE-30 on Chromosorb W
P3 = 30% DC-550 on Celite 545

REFERENCE

Kodama, R., Yano, T., Furukawa, K., Noda, K., and Ide, H., *Xenobiotica,* 6, 377, 1976.

Table GC 10
THUJYL MONOTERPENES

Column packing	P1	P2	P3
Temperature (°C)	135	100	90
Gas; Flow rate (mℓ/min)	N_2; 20	N_2; 25	N_2;10
Column			
Length (ft)	20	5	15
Diameter (in., O.D.)	0.25	0.125	0.125
Form	Coil	Coil	Coil
Material	SS	SS	SS
Detector	F1	F1	F1
Compound		$r_{3\text{-neothujanol}}$	
(+)3-Thujone	0.495	0.711	0.391
(−)3-Isothujone	0.528	0.711	4.391
n-Butanol[a]	—	—	0.600
(+)3-Neothujanol	1.00	1.00	1.00
	(84.3 min)	(15.3 min)	(29.0 min)
(−)3-Isothujanol	1.11	1.32	2.11
(+)3-Thujanol	1.19	1.32	1.70
(−)3-Neoisothujanol	1.33	1.61	2.37

[a] Added internal standard.

Column packing P1 = 15% Carbowax 20M on Celite 545 (80-100 mesh) AWS
P2 = 10% TCEP on Varaport 30 (100-120 mesh)
P3 = 15% Glycerol on Celite 545 (100-120 mesh) AWS

REFERENCE

McDonald, K. L. and Cartlidge, D. M., *J. Chromatogr. Sci.*, 9, 440, 1971.

Table GC 11
TMS — CYCLOPENTANO MONOTERPENE GLUCOSIDES

Column packing	P1	P2	P1	P3	P4
Temperature (°C)	270	270	230	215	230
Gas; Flow rate (mℓ/min)	N_2; 60	N_2; 60	N_2; 60	N_2; 60	N_2; 60
Column					
Length (m)	1.8	1.8	0.5	0.5	0.5
Diameter (mm, I.D.)	4	4	3	3	3
Form	u	u	u	u	u
Material	Glass	Glass	Glass	Glass	Glass
Detector	FI	FI	FI	FI	FI
Compound[a]			$r_{aucubin}$		
Sucrose	0.48	0.79	0.41	0.54	0.44
Aucubin	1.00 (7.12 min)	1.00 (7.65 min)	1.00 (3.50 min)	1.00 (1.32 min)	1.00 (2.70 min)
7-Deoxyloganic acid	1.37	1.18	1.40	1.88	1.65
Catalpol	1.37	1.28	1.40	2.02	1.75
7-Deoxyloganin	1.39	1.02	1.35	1.97	2.07
Monotropein	1.57	1.65	1.73	2.33	1.89
Gardenoside	1.65	1.52	1.81	2.41	1.95
Secologanin	1.75	1.16	1.85	3.67	2.45
Loganin	1.78	1.52	1.92	2.56	2.25
Scandoside	1.98	2.00	2.31	2.43	2.51
Theviridoside	2.02	1.78	2.32	2.88	2.55
Geniposide	2.15	1.65	2.42	3.17	2.89
Scandoside methyl ester	2.19	1.83	2.59	3.08	2.60
7-Dehydrologanin	2.30	1.39	2.60	6.32	5.35
Morroniside	2.53	1.96	2.91	3.63	3.37
Hastatoside	2.55	1.70	2.90	5.19	4.46
Forsythide	2.55	2.30	3.00	4.68	4.24
Forsythide 10-methyl ester	2.74	2.12	3.20	4.93	4.52
Forsythide dimethyl ester	2.75	1.75	3.20	5.07	4.56
Verbenalin	2.75	1.60	3.20	7.44	6.35
Sweroside	3.07	1.60	3.21	11.88	8.75
Gentiopicroside	3.08	1.63	3.22	9.56	8.25
Swertiamarin	3.08	1.57	3.31	9.00	7.08
Bakankosin	5.31	2.16	5.86	13.55	14.75
Kingiside	5.34	2.61	7.10	20.06	18.59
Amaroswerin	6.18	—	8.20	—	—
Amarogentin	6.21	—	8.30	—	—
Asperuloside	6.70	3.08	9.43	20.06	16.06
Paederoside	—	4.20[b]	10.80[c]	—	—
Ligustroside	—	10.00[b]	22.00[c]	—	—
Catalposide	—	12.20[b]	24.32[c]	—	—
Oleuropein	—	13.20[b]	28.40[c]	—	—
10-Acetoxyligustroside	—	15.80[b]	42.00[c]	—	—
10-Acetoxyoleuropein	—	22.00[b]	56.80[c]	—	—

[a] TMS derivatives.
[b] A 0.5 m long column was used.
[c] Column temperature was 270°C.

Column packing P1 = 1.5% OV-17 on Shimalite W AW DMCS (80/100 mesh)
P2 = 1.5% OV-1 on Shimalite W AW DMCS (80/100 mesh)
P3 = 2% OV-210 on Shimalite W AW DMCS (80/100 mesh)
P4 = 2% OV-225 on Shimalite W AW DMCS (80/100 mesh)

Table GC 11 (continued)
TMS — CYCLOPENTANO MONOTERPENE GLUCOSIDES

REFERENCE

Inouye, H., Uobe, K., Hirai, M., Masada, Y., and Hashimoto, K., *J. Chromatogr.*, 118, 201, 1976.

Table GC 12
SESQUITERPENE HYDROCARBONS

Column packing	P1	P2	P3	P4	P5	P6	P7
Temperature (°C)	155	170	130	165	132	165	160
Gas; Flow rate (mℓ/min)	He;15-40	He; 15-40	He; 15-40	He; 15-40	He; 15-40	He; 15-40	He; 15-40
Column							
Length (ft.)	16-50	16-50	16-50	16-60	16-50	16-50	16-50
Diameter (in., O.D.)	0.125	0.125	0.125	0.125	0.125	0.125	0.125
Form	na	na	na	na	na	na	na
Material	na	na	na	na	na	na	na
Detector	TC	TC	TC	TC	TC	TC	TC
Compound				I			
Cubebene	1368	—	—	—	—	—	—
α-Longipinene	—	—	1359.5	—	—	—	—
α-Ylangene	1401.5	1396.1	—	1454.5	—	1538.5	1653
β-Elemene	1410	—	—	—	—	—	—
α-Bourbonene	1410	—	—	—	—	—	—
α-Copaene	1410.2	1400.5	1378.5	1459	1447	1551.3	1665
Cyclosativene	1411.9	1399.7	—	—	—	1549	(1684)
Longicyclene	1417.1	1409.1	1371	1454	1465	1554	1684
Cyclocopacamphene	1417.8	—	—	—	1467.2	1555.4	1685.6
β-Bourbonene	1418.3	1411.7	1386	1477.3	1477.5	1586.5	1714
β-Farnesene	1429.2	—	—	—	1509	1668	1818.5
Sativene	1434.7	1420.7	—	—	—	1594.5	(1738)
Cyperene	1446.6	1432.5	1398	1501	1493	1606	1736.5
α-Gurjunene	—	1435.2	1413	1500.5	1471	1591	1712.5
Caryophyllene	1451.7	1445.3	1417.5	1523	(1587)	1655.5	1835.5
Longifolene	1464.0	1440.2	1404	1517.5	1520	1643	1802.5
Isosativene	1464.4	1440.9	—	—	—	1639	(1797)
Calarene	1466.0	1459.7	1435	1535.5	1513	1655.5	1806
β-Ylangene	—	—	1417.5	—	—	—	—
β-Copaene	—	—	1422.5	—	—	—	—
α-Cedrene	1473.4	1445.0	1414	1516	1518	1640	1788.5
Thujopsene	1476.1	1458.3	1430.5	1542.3	1540	1684.2	1858.5
Aromadendrene	1477	—	—	—	—	—	—
α-Maaliene	1477	—	—	—	—	—	—
γ-Curcumene	1481.9	—	—	—	1532.5	—	—
β-Cedrene	1482.4	1454.8	1421	1535.5	1539.5	1670	1834.5
α-Curcumene	1483	1480.4	(1475)	1589	1557.5	1787.5	1992.5
ε-Muurolene	1484.8	(1474.0)	(1445)	(1561.5)	1552.5	1713.8	1893.5
Humulene	1487.2	1476.6	1446.8	1561.5	1583.5	1719	1929.5
Santalene (minor)	—	1459.5	1441	1535	1522	1671	1830
Santalene (major)	—	1470.5	1454	1548	1533.5	1683	1843.5
Selina-4(14), 7-diene	1491.9	1475.7	—	—	1542	1694	1852.5
δ-Selinenene	1504.5	—	—	—	—	1728.5	—
γ-Muurolene	1505.7	—	—	—	1545	1725	1889

Table GC 12 (continued)
SESQUITERPENE HYDROCARBONS

Compound	I						
γ-Amorphene	1506.4	—	—	—	1544.5	1724	1896.5
α-Himachalene	1508.0	—	1444	1561.5	1533.5	1704.5	1870
α-Amorphene	1509.5	1491.7	—	1582.5	1535	1724.5	1897
Zizaene	1511.6	1481.8	—	—	1562	1706.3	1879.5
γ-Bisabolene	1512.9	1510.3	1496.5	1592.5	1548	1745.5	1909.5
β-Curcumene	1513.6	1510.4	—	—	1547.5	1756	1922.5
α-Zingaberene	—	(1479.6)	—	1583.5	—	1738	—
Valencene	1525.6	1508.8	1457	1600	1581	1760	1948
β-Himachalene	1529.7	—	1491	1607.5	1578	1752.5	—
β-Selinene	1530.2	1506.3	—	1595	1597.5	1766.5	1958
γ-Bisabolene	1531.3	(1505)	—	1601	—	1765.6	—
α-Muurolene	1531.3	1507.7	1495	1600	1558.5	1752.5	1927.5
α-Pyrovetivene	1533.9	1521.5	—	—	—	1817	2026.0
α-Selinene	1534.5	—	—	—	—	—	—
ϵ-Bulgarene	1538	—	—	—	—	—	—
δ-Cadinene	1546.4	(1526.4)	1504	1628.5	—	1784	1959
Calamenene	1550	—	—	—	—	—	—
γ-Cadinene	1554.9	1523.5	1506.5	1623.5	1587	1792.3	1978.5
Selina-4(14),7(11)-diene	1572.0	—	—	—	1611.5	1816.3	(2018)
Selina-3,7(11)-diene	1580	—	—	—	—	—	—
β-Vetivenene	1583.0	1563.3	—	—	—	1885	2111

Column packing P1 = 0.5-3% Apiezon L on silanized Chromosorb G
P2 = 0.5-3% SF-96 on silanized Chromosorb G
P3 = 0.5-3% SE-30 on silanized Chromosorb G
P4 = 0.5-3% DC-710 on silanized Chromosorb G
P5 = 0.5-3% QF-1 on silanized Chromosorb G
P6 = 0.5-3% Carbowax 20M on silanized Chromosorb G
P7 = 0.5-3% DEGS on silanized Chromosorb G

Note: Parenthetic values were determined on only one occasion. Kovat's indexes were determined for key sesquiterpene standards (α-copaene, longifolene, valencene, and γ-cadinene) using even C-numbered *n*-alkanes. All subsequent indexes for other sesquiterpenes were obtained by coinjection with a pair of sesquiterpene standards.

REFERENCE

Anderson, N. H. and Falcone, M. S., *J. Chromatogr.*, 44, 52, 1969.

Table GC 13
SESQUITERPENES

Column packing	P1	P2	P3
Temperature (°C)	90	120	60-190° at 4°/min
Gas; Flow rate (mℓ/min)	He; 50	He; 50	He; 20
Column			
Length (ft)	6	6	10
Diameter (mm)	2	2	2
Form	na	na	na
Material	Glass	Glass	Glass
Detector	FI	FI	Mass spec
Reference	1	1	2

Compound	$r_{\beta\text{-caryophyllene}}$		$r_{n\text{-heptyl acetate}}$
β-Caryophyllene	1.00	1.00	1.615 ± 0.049
trans-α-Bergamotene	1.12	0.87	—
β-Farnesene	1.26	1.23	—
Humulene	1.19	1.39	—
Elemol	—	—	2.45 ± 0.082
γ-Eudesmol	—	—	2.662 ± 0.115
α-Eudesmol	—	—	2.774 ± 0.132
β-Eudesmol	—	—	2.774 ± 0.132

Column packing P1 = 3% OV 101 on Gas Chrom Q (100-120 mesh)
P2 = 20% Reoplex 400 on Chromosorb W (80-100) AW
P3 = 1% Polyethylene glycol 20M + 1% OV 17 on Chromosorb G (80-100 mesh)

REFERENCES

1. **Hood, L. V. S., Darnes, M. E., and Barry, G. T.,** *Nature (London),* 242, 402, 1973.
2. **Adams, R. P., Granet, M., Hogge, L. R., and von Rudloff, E.,** *J. Chromatogr.,* 17, 75, 1979.

Table GC 14
HYDROGENATED SESQUITERPENES

Column packing		P1	P2
Temperature (°C)		150	130
Gas; Flow rate (mℓ/min)		na	na
Column			
Length (ft)		500	500
Diameter (in., O.D.)		0.03	0.03
Form		Capillary	Capillary
Material		SS	SS
Detector		FI	FI
Compound	**Hydrogenation Product**	**I**[a]	
Nerolidyl acetate	Farnesane	1392	1362
trans-β-Farnesene	Farnesane	1392	1360
β-Sinensal	3,7-Dimethyldodecane	1327	1340
α-Curcumene	Dihydrocurcumene	1448	1696
	Bisabolane-a	1448	1492
	Bisabolane-b	1458	1510
β-Elemene	Elemane	1404	1460
γ-Elemene	Elemane	1403	1460
Germacrene-B	Elemane	1403	1457
	Germacrane-b	1477	1572
	Germacrane-c	1482	1585
	Germacrane-d	1489	1593
	4αH,5αH-Eudesmane	1497	1636
Germacrene-D	Cadinane-a (?)	1437	—
	Eudesmane isomer	1456	—
	Eudesmane isomer	1460	—
	Germacrane-b	1475	—
	Germacrane-c	1479	—
	Unknown	1487	—
	Germacrane-d	1489	—
	4αH,5αH-Eudesmane	1495	—
Germacrone	Hexahydrogermacrone-a	1585[a]	1950[a]
	Hexahydrogermacrone-b	1590[a]	1938[a]
	Hexahydrogermacrone-c	1600[a]	1967[a]
Hexahydrogermacrone (mixture of isomers)	Unknown (m+ = 204)	1400	—
	Germacrane isomer?	1419	—
	Germacrane isomer?	1432	—
	Cadinane-a?	1443	—
	Cadinane-b?	1456	—
	Eudesmane isomer	1460	—
	Eudesmane isomer	1463	—
	Germacrane-b	1476	—
	Germacrane-c	1480	—
	Germacrane-d	1487	—
	4αH,5αH-Eudesmane	1495	—
δ-Selinene	4βH,5αH-Eudesmane	1404	1578
	4αH,5αH-Eudesmane	1497	1632
	Eudesmane isomer	1458	—
	Eudesmane isomer	1461	—
Valencene	Valencane (nootkatane)	1493	1624
Caryophyllene	Caryophyllane-a	1425	1522
	Caryophyllane-b	1432	1533
	Caryophyllane-c	1450	1555
	Caryophyllane-d	1450	1562

Table GC 14 (continued)
HYDROGENATED SESQUITERPENES

Compound	Hydrogenation Product	I[a]	
α-Cedrene	?	1449	1600
	8βH-Cedrane	1458	1617
	8αH-Cedrane	1465	1627
Cedrol	α-Cedrene	1431	—
	8βH-Cedrane	1455	1617
	8αH-Cedrane	1463	1625
Thujopsene	Tetrahydrothujopsene-a	1496	1668
	Tetrahydrothujopsene-b	1508	1678
Longifolene	?	1424	1575
	7αH-Longifolane	1460	1627
	7βH-Longifolane	1467	1633

[a] Measured relative to caryophyllene, humulene, and cuparene.

Column packing P1 = SF-96 (50)
P2 = Carbowax 20M

REFERENCE

Maarse, H., *J. Chromatogr.*, 106, 369, 1975.

Table GC 15
SESQUITERPENE DEHYDROGENATION PRODUCTS

Column packing	P1	P2
Temperature (°C)	200	195
Gas; Flow rate (mℓ/min)	He; 15-40	He; 15-40
Column		
Length (ft)	16-50	16-50
Diameter (in., O.D.)	0.125	0.125
Form	na	na
Material	na	na
Detector	TC	TC

Compound	$r_{cadelene}$[a]	I[b]	$r_{cadelene}$[a]	I[c]
Acenaphthene	1.001	2634	0.529	1590.6
Fluorene	1.746	2837	0.806	1686.4
"Sesquiterpenes"	0.08—0.21	—	0.27—0.56	—
α-Curcumene	0.201	—	0.354	1497.5
1,6-Dimethylnaphthalene	0.632	2465	0.499	1576.0
Eudalene	0.719	2512	0.624	1627.1
Cadalene	1.000	2633	1.000	1735.5
S-Guaiazulene	1.760	2840	1.550	1835.7
Chamazulene	1.793	2847	1.360	1805.8
Se-Guaiazulene	1.913	2870	1.675	1853.0
Vetivazulene	2.095	2904	1.775	1866.6

[a] = ±0.7%.
[b] = ± 2 units.
[c] = ± 0.8 units.

Column packing P1 = DEGS
P2 = Apiezon L

REFERENCE

Andersen, N. H., Falcone, M. S., and Syrdal, D. D., *Phytochemistry*, 9, 1341, 1970.

Table GC 16
SESQUITERPENES AND THEIR HYDROGENATION PRODUCTS

Column packing	P1	P2
Temperature (°C)	220	160
Gas; Flow rate (mℓ/min)	Ar; 33-43	Ar; 33-43
Column		
Length (m)	4	6
Diameter (mm)	4	3
Form	na	na
Material	na	na
Detector	F1	F1

Compound	r_I	r_{II}	r_{III}	r_{IV}	r_I	r_{II}	r_{III}	r_{IV}
trans-Nerolidol	1	1.17	1.06	0.96	1	0.95	0.67	0.51
	—	—	—	—	—	—	0.72	0.67
Fokienol	1.17	1	1.17	1.06	1.88	1.12	1.06	—
	—	—	1.17	—	—	1	0.95	0.72
trans-Nerolidyl acetate	1.47	1.66	1.45	1.30	0.94	—	—	—
Elemol	1.16	1.35	1.18	—	1.71	1.49	0.97	—
Siamol	1.46	—	1.15	—	1.89	1.13	0.83	—
	—	—	1.40	—	—	1.26	1.08	—
	—	—	—	—	—	1.49	—	—
α-Bisabolol	2.06	1.85	1.53	—	2.04	1.42	0.84	—
	—	—	1.61	—	—	—	0.93	—
T-Cadinol	1.91	1.55	—	—	1.89	1.03	—	—
	—	1.70	—	—	—	1.20	—	—
α-Cadinol	2.01	2.01	—	—	2.49	1.47	—	—
	—	—	—	—	—	1.72	—	—
γ-Eudesmol	1.80	—	—	—	1.85	—	—	—
	—	1.81	—	—	—	1.65	—	—
α-Eudesmol	1.08	—	—	—	2.65	—	—	—
	—	2.19	—	—	—	2.18	—	—
β-Eudesmol	2.08	—	—	—	2.86	—	—	—
Dauca-6(14),11-dien-5-ol	2.66	2.86	3.14	—	6.12	5.83	4.25	—
Dauca-6(14),11-dien-5-acetate	3.13	3.43	3.78	—	3.54	—	—	—
Dauca-11-en-5-one	2.38	2.71	—	—	3.78	3.04	—	—
Permoulone	—	—	—	—	3.05	1.76	1.13	0.97
δ-Cadinene	1.23	—	1.04	—	0.28	—	0.20	—

Note: I = unhydrogenated; II = dihydro; III = tetrahydro; IV = hexahydrosesquiterpene.

Column packing P1 = Apiezon L
P2 = Mannitol-hexapropionitrile ether

REFERENCE

Korthals, H.-P., Merkel, D., and Muhlstadt, M., *Ann. Chem.*, 745, 31, 1971.

Table GC 17
TRICHOTHECENE SESQUITERPENES

Column packing	P1	P1	P2
Temperature (°C)	T1	T2	T3
Gas; Flow rate (mℓ/min)	N_2; 30	N_2; 30	N_2; 30
Column			
Length (ft)	5	5	5
Diameter (in., O.D.)	1/8	1/8	1/8
Form	Coil	Coil	Coil
Material	Glass	Glass	Glass
Detector	FI	FI	FI

Compound[a]	**Methylene units**		
Trichodermol	20.3	—	20.1
Trichodermin	20.5	—	20.2
Trichothecolone	21.5	—	21.2
Crotocol	21.5	—	21.2
Verrucarol	22.3	—	22.1
Scirpenetriol	23.2	23.3	23.1
Crotocin	—	23.8	23.3
Trichothecin	—	23.8	23.4
Diacetoxyscirpenol	—	23.9	23.7
Scirpene triacetate	—	24.6	24.4
T-2 tetraol	—	24.7	24.6
T-2 toxin	—	27.8	27.7
HT-2 toxin	—	27.9	27.9

[a] Compounds having free hydroxyl groups were chromatographed as TMS ethers.

Column packing P1 = 3% SE-30 on Chromosorb W (80-100 mesh)
P2 = 3% SE-30 on Gas-Chrom Q (80-100 mesh)

Temperature T1 = 210°C
T2 = 238°C
T3 = Program 160-250°C at 4.2°/min

REFERENCE

Ikediobi, C. O., Hsu, I. C., Bamburg, J. R., and Strong, F. M., *Anal. Biochem.*, 43, 327, 1971.

Table GC 18
DITERPENES

Column packing	P1	P2	P3	P4	P5
Temperature (°C)	192	193	193	na	na
Gas; Flow rate (mℓ/min)	Ar; 45	na	na	na	na
Column					
Length	274 cm	na	na	6 ft	6 ft
Diameter	0.84 cm	na	na	0.25 in. O.D.	0.25 in. O.D.
Form	na	na	na	U	U
Material	Glass	na	na	SS	SS
Detector	FI	na	na	TC	TC
Reference	1	2	2	3	3

Compound[a]	$r_{sandaracopimarate}$			$r_{pimarate}$	
Thunbergol	0.56	—	—	—	—
epi-Manool	0.58	—	—	—	—
Pimarate	0.91	—	—	—	—
Sandaracopimarate	1.0	1.0	1.0	—	—
Laevopimarate/palustrate	1.14	—	—	—	—
Isopimarate	1.17	—	—	—	—
Dehydroabietate	1.42	1.4	1.26	—	—
Abietate	1.58	—	—	—	—
Isopimarol	1.66	—	—	—	—
epi-Torulosate	1.73	—	—	—	—
Neoabietate	1.81	—	—	—	—
epi-Torulosal	2.04	—	—	—	—
Larixol	2.07	—	—	—	—
Dehydroabietol	2.16	—	—	—	—
Abietol	2.25	—	—	—	—
Larixyl acetate	2.41	—	—	—	—
epi-Torulosal	3.08	—	—	—	—
epi-Torulosal acetate	3.08	—	—	—	—
trans-Communic acid	—	1.06	1.0	—	—
Lambertianic acid	—	1.53	1.26	—	—
Isocupressic acid	—	3.72	2.06	—	—
Isoagatholal	—	4.3	1.82	—	—
7-Oxo-dehydroabietic acid	—	5.4	2.58	—	—
Pinusolide	—	17.8	4.4	—	—
Imbricataloic acid	—	—	—	3.63	1.58
Imbricataloic acid	—	—	—	—	2.01
Artifact methyl ketone	—	—	—	4.20	1.97
Dihydroagathic acid	—	—	—	3.64	2.03

[a] Compounds refer to methyl esters where appropriate.

Column packing P1 = 1% XE-60 on diatomite CQ (100-120 mesh)
P2 = XE-60
P3 = OV-1
P4 = DEGS
P5 = SE-30/EGiP

REFERENCES

1. **Mills, J. S.,** *Phytochemistry,* 12, 2407, 1973.
2. **Gough, L. J. and Mills, J. S.,** *Phytochemistry,* 13, 1612, 1974.
3. **Spalding, B. P., Zinkel, D. F., and Roberts, D. R.,** *Phytochemistry,* 10, 3289, 1971.

Table GC 19
DITERPENE ACETATES

Column packing	P1	P2	P3	P4
Temperature (°C)	213	228	186	220
Gas; Flow rate (mℓ/min)	N_2; 60	N_2; 60	N_2; 60	na
Column				
Length	na	na	na	na
Diameter	na	na	na	na
Form	na	na	na	na
Material	na	na	na	na
Detector	na	na	na	na

Compound		$r_{codeine}$		
Ingenol triacetate	2.78	2.52	5.58	2.52
Phorbol triacetate	6.16	5.47	24.4	7.33
12-Deoxyphorbol diacetate	3.95	3.59	12.6	4.17
4-Deoxy-4α-phorbol triacetate	4.65	4.16	17.1	5.25
Crotophorbolone monoacetate	3.23	2.95	13.9	4.24
Codeine	17.0 min	29.1 min	3.65 min	12.0 min

Column packing P1 = 10% SE-30 + 0.05% EGS
P2 = 10% SE-52
P3 = 2% QF-1
P4 = 2% OV-17

REFERENCE

Kingshorn, A. D. and Evans, F. J., *J. Pharm. Pharmac.*, 26, 408, 1974.

Table GC 20
TRITERPENES

Column packing	P1	P1	P2
Temperature (°C)	240	250	255
Gas, Flow rate (mℓ/min)	He; 30	N_2; 50	N_2; 50
Column			
Length (m)	2	2	2
Diameter (I.D., mm)	3	3	3
Form	na	na	na
Material	Glass	Glass	Glass
Detector	FI	FI	FI
Reference	1	2	3

Compound		$r_{\beta\text{-sitosterol}}$ [a]	
Lophenol	—	0.83	—
Obtusifoliol (4, 14 -dimethyl-24-methylene-Δ^8-cholesten-3 -ol)	0.95	0.94	—
Cycloeucalenol(4, 14 -dimethyl-9, 19-cyclopropane-24-methylene cholestan-3β-ol)	1.10	1.11	—
Gramisterol(24-methyleneophenol) (4 -methyl-24-methylene-Δ^7-cholesten-3β-ol)	1.13	1.13	—
Citrostadienol(4 -methyl- 24Z -24-ethylidene-Δ^7-cholesten-3β-ol)	1.52	1.52	—
24-Dihydroparkeol	—	1.01	—
Cycloartanol	1.02	1.02	1.02
β-Amyrin	1.13	1.13	—
Cycloartenol	1.23	1.24	1.23
α-Amyrin	1.29	1.28	—
24-Methylenecycloartanol	1.38	1.38	1.37
Cyclobranol(24-methylcycloartenol)	1.68	—	—
Cholesterol	—	0.61	—
Brassicasterol	—	0.70	—
Campesterol	—	0.81	—
Stigmasterol	—	0.88	—
24-Methylcholest-7-enol	—	0.95	—
β-Sitosterol	—	1.00	—
α-Spinasterol	—	1.03	—
Δ^5-Avenasterol	—	1.12	—
Δ^7-Stigmasterol	—	1.18	—
Δ^7-Avenasterol	—	1.32	—
Parkeol	—	1.22	1.22
24-Methylenelanost-9(11)-enol	—	1.37	1.35
Butyrospermol	—	1.17	—
Lupeol	—	1.33	—
24-Methylenecycloartanol acetate	—	—	1.69
24-Methylcycloartanol	—	—	1.33
24-Methylcycloartanol acetate	—	—	1.64
Cycloartenol acetate	—	—	1.52
Cycloartanol acetate	—	—	1.26
24-Methylenelanost-9(11)-enol acetate	—	—	1.62
24-Methyllanost-9(11)-enol	—	—	1.31
24-Methyllanost-9(11)-enol acetate	—	—	1.57
Parkeol acetate	—	—	1.46
Lanost-9(11)-enol	—	—	1.01
Lanost-9(11)-enol acetate	—	—	1.21
24-methyllanost-8-enol	—	—	1.15
24-Methyllanost-8-enol acetate	—	—	1.39
Lanosterol	—	—	1.07
Lanosterol acetate	—	—	1.28

Table GC 20 (continued)
TRITERPENES

Compound	$r_{\beta\text{-sitosterol}}$ [a]		
Lanost-8-enol	—	—	0.89
Lanost-8-enol acetate	—	—	1.07
24-Methyllanost-7-enol	—	—	1.38
24-Methyllanost-7-enol acetate	—	—	1.67
Lanost-7-enol	—	—	1.06
Lanost-7-enol acetate	—	—	1.28

[a] t_r β-sitosterol = 30 min under all three conditions.

Column packing P1 = 1.5%
P2 = 3% OV-17 on Gas Chrom-Z (80-100 mesh)

REFERENCES

1. **Itoh, T., Tamura, T., and Matsumoto, T.,** *J. Am. Oil Chem. Soc.*, 50, 300, 1973.
2. **Itoh, T., Tamura, T., and Matsumoto, T.,** *Lipids*, 9, 173, 1974.
3. **Itoh, T., Tamura, T., and Matsumoto, T.,** *Lipids*, 10, 454, 1975.

Table GC 21
TRITERPENOLS

Column packing	P1	P2	P3
Temperature (°C)	240	270	294
Gas; Flow rate (mℓ/min)	na	N_2; 0.5	N_2; 30
Column			
Length	1.20 m	25 m	5 ft
Diameter (mm, I.D.)	4	0.24	4
Form	na	Capillary	na
Material	Glass	na	Glass
Detector	na	na	FI
Reference	1	2	3

Compound	$r_{5\alpha\text{-cholestane}}$		$r_{cholesterol}$
Lupeol	3.31	3.01	4.30[a]
α-Amyrin	2.97	2.76	1.96[a]
Cycloartenol	3.35	—	—
β-Amyrin	3.27	3.03	2.15[b]
Germanicol	3.07	2.84	—
Cyclolaudenol	3.92	—	—
Taraxerol	2.84	—	—
Taraxasterol	4.01	—	6.21[a]
24-Methylene-cycloartanol	4.04	—	—
Bauerenol	3.90	—	—
Lanosterol	3.01	—	—
ψ-taraxasterol	—	—	6.21
Erthrodiol	—	—	3.00
Erthrodiol acetate	—	—	3.39
Brein	—	—	3.44
Brein acetate	—	—	3.53
Ursadiol	—	—	3.44
Ursadiol acetate	—	—	3.78
Calenduladiol	—	—	4.05
Calenduladiol acetate	—	—	4.40
Faradiol	—	—	5.09
Faradiol acetate	—	—	5.66

[a] Relative retention time of selenium dioxide oxidation product.
[b] Not oxidized.

Column packing P1 = 3% SE-30 on Varaport 30
P2 = SE-30
P3 = 1.5% OV-17 on Shimalite W

REFERENCES

1. **Nieman, G. J. and Baas, W. J.,** *J. Chromatogr. Sci.*, 16, 260, 1978.
2. **Baas, W. J.,** *J. Chromatogr.*, 153, 263, 1978.
3. **Wilkomirski, B. and Kasprzyk, Z.,** *J. Chromatogr.*, 129, 440, 1976.

Table GC 22
TRITERPENE AND STEROL ACETATES

Column packing	P1	P2	P3	P4
Temperature (°C)	225	205	225	220
Gas; Flow rate (mℓ/min)	N_2 ;40	N_2 ;25	N_2 ;25	N_2 ;32
Column				
Length (m)	2	2	2	2
Diameter (mm)	1.5	1.5	1.5	1.5
Form	na	na	na	na
Material	Glass	Glass	Glass	Glass
Detector	ms	ms	ms	ms
Compound[a]		t_r		
Cholesterol (1.00)	1.00	1.00	1.00	1.00
Campesterol (1.28)	1.28	1.28	1.30	1.33
Stigmasterol (1.41)	1.41	1.29	1.40	1.33
Sitosterol (1.63)	1.63	1.60	1.68	1.59
Sterol A	1.10	1.03	1.08	—
24-Methylenecholesterol	1.31	1.28	1.25	—
24-Methylenecholest-7-en-3β-ol	1.64	—	1.41	—
Sterol B	—	—	1.54	—
28-Isofucosterol (1.80)	1.80	1.62	1.66	1.83
24-Ethylidenecholest-7-en-3β-ol (1.91)	1.91	1.73	1.86	2.10

[a] Separated as acetates.

Column packing P1 = 3% OV-17; RR_t 5α-cholestane to cholesterol and cholesteryl acetate, 0.37 and 0.26, respectively.
P2 = 2.5% QF-1; RR_t 5α-cholestane to cholesteryl acetate, 0.21
P3 = 1% SE-33; RR_t 5α-cholestane to cholesteryl acetate, 0.34
P4 = 1% Hieff 8B + 2% PVP

REFERENCE

Palmer, M. A. and Bowden, B. N., *Phytochemistry*, 14, 2049, 1975.

Table GC 23
STEROLS

Column packing	P1
Temperature (°C)	270
Gas; Flow rate (mℓ/min)	He ;60
Column	
Length (cm)	240
Diameter (mm, I.D.)	3
Form	Coil
Material	Glass
Detector	FI

Compound	t_r	$r_{diosgenin}$
25R-Spirosta-3,5-diene	4.3	0.64
5α-Cholestan-3β-ol	5.0	0.75
Smilagenin	6.2	0.93
Diosgenin	6.7	1.00
Yamogenin	6.9	1.03
Tigogenin	7.0	1.05
Pennogenin	9.2	1.37

Column packing P1 = 2% SE-30 on Gas Chrom Q (100—120 mesh)

REFERENCE

Rozanski, A., *Analyst*, 97, 968, 1972.

Table GC 24
ECDYSONE TRITERPENES

Column packing	P1	P1
Temperature (°C)	275	260
Gas; Flow rate (mℓ/min)	N_2 ;40	N_2 ;40
Column		
Length (cm)	150	150
Diameter (in., O.D.)	4	4
Form	U	U
Material	Glass	Glass
Detector	FI	FI

Compound	$r_{cholesterol\ butyrate}$ TMS	Partial TMS	Heptafluorobutyrate
2β,3β,14α-Trihydroxy-5β-cholest-7-en-6-one	1.37 (Tri)	—	0.61
α-Ecdysone	2.62 (Penta)	—	1.07
Ponasterone A	2.33 (Penta)	2.23 (Tetra)	1.00
Ecdysterone	3.61 (Hexa)	3.12 (Penta)	1.86
Inokosterone	4.24 (Hexa)	3.73 (Penta)	1.49
Makisterone A	4.45 (Hexa)	4.07 (Penta)	2.27
Makisterone B	4.28 (Hexa)	3.62 (Penta)	1.52
Cyasterone	10.09 (Penta)	9.54 (Tetra)	4.94
Cholesteryl Butyrate	1.00[a]	—	1.00[b]

[a] t_r = 3.3 min.
[b] t_r = 6.1 min.

Column packing P1 = 1.5% OV-101 on acid-washed and silanized Chromosorb W (80-100 mesh)

REFERENCE

Ikekawa, N., Hattori, F., Rubio-Lightbourn, J., Miyazaki, H., Ishibashi, M., and Mori, C., *J. Chromatogr. Sci.*, 10, 233, 1972.

Table GC 25
POLYPRENOLS

Column packing	P1	P1
Temperature (°C)	T1	T2
Gas; Flow rate (mℓ/min)	Ar; 60	Ar; 60
Column		
Length (ft)	4	4
Diameter (in., O.D.)	0.125	0.125
Form	na	na
Material	SS	SS
Detector	FI	FI
References	1, 2	1, 2

Compound	t_r (min)	
Perhydroacetates of		
Betulaprenol-6	0.48	0.29
Betulaprenol-7	1.28	0.47
Betulaprenol-8	2.86	0.82
Betulaprenol-9	6.47	1.47
Solanesol	6.47	1.47
Castaprenol-10	—	2.80
Castaprenol-11	32.2	5.48
Castaprenol-12	69.0	10.0
Castaprenol-13	—	18.20
Ficaprenol-9	6.0	1.48
Ficaprenol-10	13.6	2.68
Ficaprenol-11	29.7	5.44
Ficaprenol-12	62.1	9.94
Ficaprenol-13	—	17.8
Acetates of		
Betulaprenol-6	0.57	—
Betulaprenol-7	1.10	—
Betulaprenol-8	2.15	—
Solanesol	5.85	—
Castaprenol-10	—	1.85
Castaprenol-11	22.0	3.82
Castaprenol-12	47.2	7.40
Castaprenol-13	—	13.10
Ficaprenol-10	10.0	—
Ficaprenol-11	22.4	—
Saturated hydrocarbons of		
Solanesol	—	0.78
Castaprenol-10	—	1.42
Castaprenol-11	—	2.80
Castaprenol-12	—	5.42
Castaprenol-13	—	7.50
Ficaprenol-9	—	0.75
Ficaprenol-10	—	1.42
Ficaprenol-11	—	2.81
Ficaprenol-12	—	5.33

Column packing P1 = 1% SE-30 on silanized Chromosorb W

Temperature T1 = 300°C
T2 = 340°C

Table GC 25 (continued)
POLYPRENOLS

REFERENCES

1. **Wellburn, A. R., Stevenson, J., Hemming, F. W., and Morton, R. A.,** *Biochem. J.*, 102, 313, 1967.
2. **Stone, K. J., Wellburn, A. R., Hemming, F. W., and Pennock, J. F.,** *Biochem. J.*, 102, 325, 1967.

Table GC 26
ISOPRENOID ALCOHOLS

Column packing	P1
Temperature (°C)	110-300 at 3.71°C/min
Gas; Flow rate (mℓ/min)	N_2; 40
Column	
Length (cm)	168
Diameter (mm, I.D.)	6.5
Form	na
Material	Glass
Detector	FI

	TRE[a]		
Compound	**Free alcohol**	**TMS ether**	**Acetate**
Geraniol	0.492 ± 0.006	0.527 ± 0.002	0.538 ± 0.003
cis-trans-Nerolidol	0.591 ± 0.009	0.579 ± 0.003	0.589 ± 0.002
all-*trans*-Nerolidol	0.603 ± 0.006	0.599 ± 0.004	0.603 ± 0.002
cis-trans-Farnesol	0.654 ± 0.005	0.682 ± 0.004	0.691 ± 0.003
all-*trans*-Farnesol	0.662 ± 0.005	0.695 ± 0.001	0.708 ± 0.003
Phytol	0.802 ± 0.003	0.823 ± 0.001	0.835 ± 0.002
cis-trans-Geranyl geraniol	0.809 ± 0.002	0.834 ± 0.001	0.841 ± 0.001
all-*trans*-Geranyl geraniol	0.824 ± 0.005	0.845 ± 0.002	0.851 ± 0.002

[a] TRE = Relative elution temperature (squalene = 1.00).

Column packing P1 = 3% SE-30 on Gas Chrom Q (100-200 mesh)

REFERENCE

Watts, R. B. and Kekwick, R. G. O., *J. Chromatogr.*, 88, 15, 1974.

Table GC 27
ISOPRENOID ALKANES

Column packing	P1		P2		P3	
Temperature (°C)	200		200		200	
Gas; Flow rate (mℓ/min)	na		na		na	
Column						
Length (ft)	Variable		Variable		Variable	
Diameter (in., O.D.)	na		na		na	
Form	Capillary		Capillary		Capillary	
Material	Copper, SS		Copper, SS		Copper, SS	
Detector	FI		FI		FI	

Compound	**I**	**ΔI/ΔT**	**I**	**ΔI/ΔT**	**I**	**ΔI/ΔT**
2,6-Dimethylnonane C_{11}(2,6)	1018.5	+0.045	1026.0	+0.53	—	—
2,6-Dimethyldecane C_{12}(2,6)	1112.0	+0.070	1119.1	+0.058	—	—
2,6-Dimethylundecane C_{13}(2,6)	1207.0	+0.041	1215.5	+0.049	—	—
2,6,10-Trimethylundecane C_{14}(2,6,10)	1260.4	+0.013	1274.8	+0.039	1231.2	−0.070
2,6,10-Trimethyldodecane C_{15}(2,6,10)	1366.3	+0.030	1378.9	+0.050	1346.7	−0.049
2,6,10-Trimethyltridecane C_{16}(2,6,10)	1448.8	+0.011	1462.8	+0.031	1422.8	−0.079
2,6,10-Trimethyltetradecane C_{17}(2,6,10)	1540.0	+0.013	1554.6	+0.044	1509.9	−0.103
2,6,10-Trimethylpentadecane C_{18}(2,6,10)	1632.7	0.0	1650.2	+0.052	1602.5	−0.112
2,6,10,14-Tetramethylpentadecane C_{19}(2,6,10,14)	1686.6	0.0	1709.4	+0.028	1645.9	−0.137
2,6,10,14-Tetramethylhexadecane C_{20}(2,6,10,14)	1790.9	+0.014	1813.6	+0.052	1759.3	−0.121

Column packing P1 = Apiezon
P2 = SE-30
P3 = Diethylene glycol adipate ester cross-linked with pentaerythritol

REFERENCE

Shlyakhov, A. F., Koreshkova, R. I., and Telkova, M. S., *J. Chromatogr.*, 104, 337, 1975.

Table GC 28
HYDROGENATED CAROTENOIDS AND OTHER TERPENOIDS

Column packing	P1	P2	P3
Temperature (°C)	T1	T1	T1
Gas; Flow rate (mℓ/min)	N_2; 60	N_2; 60	N_2; 60
Column			
Length (ft)	5	5	5
Diameter (in., O.D.)	0.25	0.25	0.25
Form	na	na	na
Material	Glass	Glass	Glass
Detector	FI	FI	FI

Compound		$r_{cholestane}$	
Geranyllinalool (C_{20})	0.15	0.16	0.13
Phytol (C_{20})	0.18	0.17	0.10
Geranylgeraniol (C_{20})	0.21	0.22	0.15
Squalene (C_{30})	1.00	1.00	1.00
Cholesterol (C_{27})	1.69	1.46	1.67
Ergosterol (C_{28})	1.98	1.65	1.80
Stigmasterol (C_{29})	2.14	1.82	1.95
Lanosterol (C_{30})	2.31	1.91	2.11
Lycopersene (C_{40})	3.95	3.15	2.76
Hydrogenation products of:			
Retinol (C_{20})	0.11	0.13	0.07
Retinaldehyde (C_{20})	0.10	0.10	0.06
Crocetin (C_{20})	0.13	0.11	0.13
Dimethylcrocetin (C_{22})	0.34	0.33	0.40
Diethylcrocetin (C_{24})	0.45	0.45	0.50
Bixin (C_{25})	1.14	1.09	1.20
Methylbixin (C_{26})	1.08	1.08	1.18
β-Apo-10′-carotenal (C_{27})	0.42	0.52	0.41
Azafrin (C_{27})	2.08	1.75	1.80
Methylazafrin (C_{28})	1.70	1.49	1.73
Squalene (i.e., squalane, C_{30})	0.66	0.73	0.54
4,4′-Diapophytoene (C_{30})	0.67	0.73	0.53
4,4′-Diapophytofluene (C_{30})	0.67	0.74	0.54
4,4′-Diapo-ζ-carotene (C_{30})	0.67	0.73	0.53
4,4′-Diaponeurosporene (C_{30})	0.67	0.73	0.53
β-Apo-8′-carotenal (C_{30})	0.82	0.85	0.70
3,4-Dehydro-β-apo-8′-carotenal (C_{30})	0.82	0.85	0.69
β-Apo-8′-carotenoic acid (C_{30})	1.57	1.48	1.40
β-Apo-8′-carotenoic acid methyl ester (C_{31})	1.61	1.54	1.46
β-Apo-8′-carotenoic acid ethyl ester (C_{32})	1.77	1.59	1.50
β-Apo-4′-carotenal (C_{35})	2.10	1.84	1.44
Lycopersene (i.e., lycopersane, C_{40})	3.15	2.67	1.76
Phytoene (C_{40})	3.14	2.64	1.77
Phytofluene (C_{40})	3.15	2.66	1.75
ζ-Carotene (C_{40})	3.13	2.66	1.78
Neurosporene (C_{40})	3.14	2.68	1.77
Lycopene (C_{40})	3.15	2.66	1.75
γ-Carotene (C_{40})	3.66	2.83	2.24
β-Zeacarotene (C_{40})	3.66	2.83	2.24
β-Carotene (C_{40})	3.81	2.88	2.43
α-Carotene (C_{40})	3.81	2.87	2.43
Dehydro-β-carotene (C_{40})	3.81	2.88	2.43
Carotinin (C_{40})	3.81	2.88	2.42
Echinenone (C_{40})	4.65	3.59	3.19
Canthaxanthin (C_{40})	5.90	4.20	3.66
β-Carotenone (C_{40})	5.33	4.12	3.09

Table GC 28 (continued)
HYDROGENATED CAROTENOIDS AND OTHER TERPENOIDS

Compound	$r_{cholestane}$		
Torularhodin (C_{40})	4.23	2.99	2.58
Capsanthin (C_{40})	5.13	3.22	2.81
Astacene (C_{40})	5.79	3.58	3.00
Physalien (C_{40} + $2C_{16}$); C_{40} fragment;	3.90	2.92	2.42
acyl fragment?	0.17	0.14	0.13
Cryptoxanthin (C_{40})	4.14	3.33	2.55
Cryptoxanthin, Ac	4.10	3.28	2.53
Cryptoxanthin, TMS	4.12	3.31	2.54
Isocryptoxanthin (C_{40})	3.92	2.91	2.53
Isocryptoxanthin, Ac	3.78	2.81	2.44
Isocryptoxanthin, TMS	4.07	3.02	2.51
Zeaxanthin (C_{40})	4.68	3.45	3.30
Zeaxanthin, diAc	5.04	3.50	2.98
Zeaxanthin, diTMS	5.07	3.55	3.05
Isozeaxanthin, (C_{40})	4.16	3.00	2.59
Isozeaxanthin, diAc	4.31	3.14	2.47
Isozeaxanthin, diTMS	4.35	3.17	2.53
Dimethoxyzeaxanthin (C_{42})	5.15	3.67	3.44
Dimethoxyisozeaxanthin (C_{42})	4.02	2.88	2.37
Fucoxanthin (C_{42})	6.24	3.49	3.42
Decapreno-β-carotene (C_{50})	7.54	5.58	5.32
Retention time (min) of:			
Squalene	5.50	8.65	11.15
Perhydro-β-carotene	20.95	24.91	27.09

Column packing P1 = 2% SE-52 on Gas Chrom Q (80-100 mesh)
P2 = 2% Dow Corning high vacuum grease on Chromosorb W AWDMCS (85-100 mesh)
P3 = 3% OV-17 on Universal B (85-100 mesh)
Temperature T1 = 225-300°C at 3°/min

REFERENCE

Taylor, R. F. and Davies, B. H., *J. Chromatogr.*, 103, 327, 1975.

Table GC 29
CAROTENOIDS

Column packing	P1	P1	P1
Temperature (°C)	T1	T2	T3
Gas; Flow rate (mℓ/min)	N_2 ;75	N_2 ;75	N_2 ;75
Column			
Length (ft)	6	6	6
Diameter (mm)	4	4	4
Form	U	U	U
Material	Glass	Glass	Glass
Detector	FI	FI	FI
Compound		$r_{squalene}$	
Squalene	1.00[a]	1.00[b]	1.00[c]
Phytoene	5.15	4.46	2.04
Lycopene	0.51	0.59	0.61
Echinenone	0.55	0.62	0.62
Hydrogenated derivatives of			
Phytol	0.16	0.24	0.11
Retinol	0.19	0.24	0.13
Squalene	0.75	0.77	0.82
β-apo-8′-Carotenal	0.92	0.91	0.91
Lycopene	4.66	3.81	1.93
β-Carotene	7.24	5.98	2.33
Echinenone	7.23	5.93	2.33
Canthaxanthin	7.23 and 11.8	5.93 and 10.3	2.33 and 2.95

[a] t_r = 5.88 min.
[b] t_r = 3.88 min.
[c] t_r = 21.5 min.

Column packing P1 = 5% Dow-Corning high vacuum grease on acid-washed Chromosorb W (80-100 mesh)

Temperature T1 = 275°C
T2 = 290°C
T3 = Program 225-290°C increased 2°/min after initial isothermal period of 3 min

REFERENCE

Taylor, R. F. and Ikawa, M., *Anal. Biochem.*, 44, 623, 1971.

Table GC 30
E VITAMERS

Column packing	P1	P2	P3	P4	P5	P6
Temperature (°C)	235	235	235	270	245	235
Gas; Flow rate (mℓ/min)	He; 30	He; 30	He; 30	He; 75	He; 75	na; 30
Column						
Length (ft)	15	15	15	8	8	6
Diameter (in., O.D.)	0.125	0.125	0.125	0.25	0.25	0.25
Form	na	na	na	na	na	na
Material	Glass	Glass	Glass	Glass	Glass	Glass
Detector	FI	FI	FI	FI	FI	FI
Reference	1	1	1	2	2	3

Compound	$r_{didecyl\ pimelate}$			$r_{trioctanoin}$		$r_{\alpha\text{-tocopherol acetate isomer}}$
TMS ether of:						
Tocol	0.69	0.67	0.55	—	—	—
5-Methyltocol	0.91	0.93	0.74	—	—	—
7-Methyltocol	0.79	0.79	0.62	—	—	—
8-Methyltocol (δ-tocopherol)	0.73	0.68	0.56	—	—	—
5,7-Dimethyltocol	1.16	1.33	1.01	—	—	—
5,8-Dimethyltocol (β-tocopherol)	0.93	0.91	0.71	—	—	—
7,8-Dimethyltocol (γ-tocopherol)	0.95	0.93	0.73	—	—	—
5,7,8-Trimethyltocol (α-tocopherol)	1.37	1.53	1.14	—	—	—
Tocotrienol	0.91	0.89	0.98	—	—	—
5-Methyltocotrienol	1.21	1.26	1.32	—	—	—
7-Methyltocotrienol	1.05	1.07	1.12	—	—	—
8-Methyltocotrienol (δ-tocotrienol)	0.95	0.90	1.00	—	—	—
5,7-Dimethyltocotrienol	1.55	1.76	1.78	—	—	—
5,8-Dimethyltocotrienol (β-tocotrienol)	1.21	1.19	1.27	—	—	—
7,8-Dimethyltocotrienol (γ-tocotrienol)	1.25	1.24	1.33	—	—	—
5,7,8-Trimethyltocotrienol (α-tocotrienol)	1.81	2.04	2.05	—	—	—
Octacosane	0.57	0.66	0.29	—	—	—
Didecyl pimelate	50 min	26 min	64 min	—	—	—
Trioctanoin	—	—	—	1.00	1.00	—
Propionate ester of:						
α-Tocopherol	—	—	—	2.38	2.38	—
β-Tocopherol	—	—	—	1.96	1.96	—
γ-Tocopherol	—	—	—	1.96	1.96	—
δ-Tocopherol	—	—	—	1.56	1.54	—
α-Tocopherol acetate isomer	—	—	—	—	—	1.00
Methoxy-2,3-dimethyl-4-acetoxy-5-phytyl-benzene	—	—	—	—	—	0.97
1-Methoxy-2,3-dimethyl-4-acetoxy-6-phytyl-benzene	—	—	—	—	—	0.86
1-Methoxy-2,6-dimethyl-4-acetoxy-5-phytyl-benzene	—	—	—	—	—	0.93
1-Methoxy-3,6-dimethyl-4-acetoxy-2-phytyl-benzene	—	—	—	—	—	0.81

Table GC 30 (continued)
E VITAMERS

Compound	$r_{didecyl\ pimelate}$			$r_{trioctanoin}$		$r_{\alpha\text{-tocopherol acetate isomer}}$
1-Methoxy-3,5-dimethyl-4-acetoxy -6-phytyl-benzene	—	—	—	—	—	0.93
1-Methoxy-2,6-dimethyl-4-acetoxy -5-phytyl-benzene	—	—	—	—	—	0.88

Column packing P1 = 1.7% SE-30 on Gas Chrom Q (100-200 mesh)
P2 = 0.4% Apiezon L on Gas Chrom Q (100-200 mesh)
P3 = 1.6% OV-17 on Gas Chrom Q (100-200 mesh)
P4 = 2% SE-52 on propionated Diatoport-S (80-100 mesh)
P5 = 4% SE-30 on propionated Diatoport-S (80-100 mesh)
P6 = 3% OV-1 on Gas Chrom S (80-100 mesh)

REFERENCES

1. **Slover, H. T., Lehmann, J., and Valis, R. J.,** *J. Am. Oil Chem. Soc.*, 46, 417, 1969.
2. **Feeter, D. K., Jacobs, M. F., and Rawlings, H. W.,** *J. Pharm. Sci.*, 60, 913, 1971.
3. **Vance, J. and Bentley, R.,** *Bioorgan. Chem.*, 1, 329, 1971.

Table GC 31
TOCOPHEROLS AND TOCOTRIENOLS

Column packing	P1	P2
Temperature (°C)	235	240
Gas; Flow rate (mℓ/min)	N_2; 45	He; 80
Column		
Length	100 cm	6 ft
Diameter	0.4 cm	0.12 in
Form	na	na
Material	Glass	SS
Detector	FI	FI
Reference	1	2

Compound	$r_{\beta\text{-sitosterol}}$	$r_{5,7\text{-dimethyltocol}}$
α-Tocopherol	0.67	1.18
β-Tocopherol	0.53	0.81
γ-Tocopherol	0.53	0.82
δ-Tocopherol	0.39	0.64
α-Tocotrienol	1.06	—
β,γ-Tocotrienol	0.84	—
δ-Tocotrienol	0.62	—

Column packing P1 = 2% Silicone oil MS 550/HP on Chromosorb W (80-100 mesh) AW-DMCS
P2 = 3% OV-1 on Supelcoport (80-100 mesh)

REFERENCES

1. **Meijboom, P. W. and Jongenotler, G. A.,** *J. Am. Oil Chem. Soc.*, 56, 33, 1979.
2. **Lovelady, H. G.,** *J. Chromatogr.*, 85, 81, 1973.

Table GC 32
ISOPRENOID QUINONES

Column packing	P1
Temperature (°C)	260
Gas; Flow rate (mℓ/min)	He; 25
Column	
Length (cm)	122
Diameter (mm, I.D.)	3
Form	U
Material	Glass
Detector	FI

Compound	t_r (min)	$r_{tocopherol}$
α-Tocopherol	7.00	1.00
Phytyl ubiquinone	7.66	1.10
Phylloquinone	11.33	1.62
Menaquinone	13.66	1.95

Column packing P1 = 3.8% Silcone gum rubber SGR UC-W-982 on Diatoport (80-100 mesh)

REFERENCE

Dialameh, G. H. and Olsen, R. E., *Anal. Biochem.*, 32, 263, 1969.

Table LC 1
MONOTERPENES

Column packing	P1	P1
Solvent	S1	S2
Temperature	rt	rt
Flow rate (mℓ/min)	2	1
Column		
Length (cm)	30	30
Diameter (mm)	3.9	3.9
Material	SS	SS
Detection	D1	D1
Reference	1	2

Compound	$\mathbf{V_R}$	$\mathbf{t_r}$ **(min)**
Geranyl-3,5-dinitrobenzoate	23.6	—
Neryl-3,5-dinitrobenzoate	23.5	—
10-Hydroxygeranyl-*bis*-(3,5-dinitrobenzoate)	29.4	—
10-Hydroxyneryl-*bis*-(3,5-dinitrobenzoate)	29.7	—
Loganic acid	—	6
Secologanic acid	—	7.2
Loganin	—	12

Column packing P1 = μBondapak C_{18} reverse phase (10 μm, Waters)

Solvent S1 = 1% Acetic acid in acetonitrile-water (1:1)

S2 = Methanol-water (4:6)

Detection D1 = UV at 254 nm

REFERENCES

1. **Licht, H. J.,** Ph.D. Thesis, Department of Biochemistry, St. Louis University, St. Louis, Mo., 1979.
2. **Coscia, C. J.,** unpublished data.

Table LC 2
CONJUGATED GIBBERELLINS

Column packing	P1	P1	P1	P1	P1	P1	P1	P1	P1	P1	P1	P1	P1	P1	P2	P2	P2	P2	P2	P2
Solvent	S1	S2	S3	S4	S5	S6	S7	S8	S9	S10	S11	S12	S13	S14	S15	S16	S17	S18	S19	S20
Temperature	rt	rt	rt	rt	rt	rt	rt	rt	rt	rt	rt	rt	rt	rt	rt	rt	rt	rt	rt	rt
Flow rate (mℓ/min)	1.5	1	1	1	1	1	0.5	1	1	1	1	0.5	1	1	0.7	0.7	0.7	0.7	0.7	0.7
Column																				
Length (cm)	25	25	25	25	25	25	25	25	25	25	25	25	25	25	25	25	25	25	25	25
Diameter (mm)	2	2	2	2	2	2	2	2	2	2	2	2	2	2	2	2	2	2	2	2
Material	SS	SS	SS	SS	SS	SS	SS	SS	SS	SS	SS	SS	SS	SS	SS	SS	SS	SS	SS	SS
Detection	UV	UV	UV	UV	UV	UV	UV	UV	UV	UV	UV	UV	UV	UV	UV	UV	UV	UV	UV	UV
Compound	t_r (min)																			
A_1-3-G	—	19.0	—	4.7	—	—	—	—	—	—	—	—	—	—	—	—	—	—	—	
A_1-13-G	—	12.4	—	3.4	—	—	—	—	—	—	—	—	—	—	—	—	—	—	—	—
A_3-3-G	—	16.8	13.3	4.1	2.7	—	—	19.0	3.3	—	—	—	2.0	4.9	5.0	2.6	2.1	4.2	—	—
A_8-2-G	17.5	4.5	4.0	1.7	1.3	—	—	4.2	1.2	—	—	—	1.0	2.0	2.6	2.0	—	1.8	1.6	1.6
A_{26}-2-G	—	—	10.5	—	2.4	—	—	6.0	1.5	—	—	—	1.3	4.7	4.7	2.5	—	2.0	—	—
A_{29}-2-G	16.8	4.2	3.6	1.6	1.3	—	—	3.3	1.1	—	—	—	1.0	2.0	2.6	2.0	2.0	1.6	—	—
A_{35}-11-G	—	27.0	21.2	5.8	3.9	1.0	2.0	10.5	2.2	—	—	1.9	1.9	7.9	5.6	2.9	—	2.4	1.6	1.6
Gibb. G	—	9.5	7.6	—	—	—	—	4.9	1.4	—	—	—	1.0	2.5	1.7	—	—	1.8	—	—
A_1GE	—	—	—	—	3.9	1.0	2.0	—	15.2	2.8	1.0	1.9	7.3	7.5	—	—	2.0	—	3.2	3.0
A_3GE	—	—	—	—	3.4	1.0	1.9	—	13.0	2.3	0.8	1.8	6.3	6.5	—	—	2.0	—	3.2	3.0
A_4GE	—	—	—	—	>25.0	2.2	4.5	—	>25.0	18.0	2.2	4.6	>25.0	>25.0	—	—	2.8	—	16.5	8.0
A_{37}GE	—	—	—	—	>25.0	1.8	3.2	—	>25.0	12.2	1.9	3.2	>25.0	>25.0	—	—	2.6	—	10.0	6.2
A_{38}GE	—	—	—	—	3.3	1.0	1.9	—	13.0	2.3	0.8	1.8	5.0	6.1	—	—	2.0	—	3.2	3.0
A_1	—	—	—	—	—	—	—	—	—	—	—	—	2.2	8.9	—	3.2	2.2	—	—	1.8
A_3	—	—	—	—	—	—	—	—	—	—	—	—	2.2	7.8	—	2.8	2.2	—	—	1.8
A_4	—	—	—	—	—	—	—	—	—	—	—	—	>25.0	—	—	—	—	—	—	—
A_5	—	—	—	—	—	—	—	—	—	—	—	—	4.4	—	—	5.2	2.7	—	—	2.2
A_8	—	—	—	—	—	—	—	—	—	—	—	—	1.0	2.0	—	2.4	2.2	—	—	1.8
A_{27}	—	—	—	—	—	—	—	—	—	—	—	—	6.0	—	—	5.4	2.7	—	—	—
A_{36}	—	—	—	—	—	—	—	—	—	—	—	—	5.5	—	—	7.2	2.9	—	—	—

Column packing	P1	= Octadecylsilanized silica gel (Wako gel LC, ODS-10H 10 μm)
	P2	= Silanized Merksorb S160.
Solvent	S1	= 10 m*M* NH_4Cl (pH 3.2)
	S2	= 10% Methanol in 10 m*M* NH_4Cl (pH 3.2)
	S3	= 15% Methanol in 10 m*M* NH_4Cl (pH 3.2)
	S4	= 20% Methanol in 10 m*M* NH_4Cl (pH 3.2)
	S5	= 30% Methanol in 10 m*M* NH_4Cl (pH 3.2)
	S6	= 50% Methanol in 10 m*M* NH_4Cl (pH 3.2)
	S7	= 50% Methanol in 10 m*M* NH_4Cl (pH 3.2)
	S8	= 5% Methanol in 10 m*M* NH_4Cl (pH 6.0)
	S9	= 15% Methanol in 10 m*M* NH_4Cl (pH 6.0)
	S10	= 30% Methanol in 10 m*M* NH_4Cl (pH 6.0)
	S11	= 50% Methanol in 10 m*M* NH_4Cl (pH 6.0)
	S12	= 50% Methanol in 10 m*M* NH_4Cl (pH 6.0)
	S13	= 20% Methanol in 10 m*M* NH_4Cl (pH 5.6)
	S14	= 20% Methanol in 10 m*M* NH_4Cl (pH 3.2)
	S15	= 10% Methanol in 10 m*M* NH_4Cl (pH 3.2)
	S16	= 30% Methanol in 10 m*M* NH_4Cl (pH 3.2)
	S17	= 50% Methanol in 10 m*M* NH_4Cl (pH 3.2)
	S18	= 10 m*M* NH_4Cl (pH 5.5)
	S19	= 20% Methanol in 10 m*M* NH_4Cl (pH 5.5)
	S20	= 30% Methanol in 10 m*M* NH_4Cl (pH 5.5)
Detection	UV 200-210 nm (end absorption)	

REFERENCE

Yamaguchi, I., Yokota, T., Yoshida, S., and Takahashi, N., *Phytochemistry,* 18, 1699, 1979.

Table LC 4
TRITERPENES

Column packing	P1		P1	P1
Solvent	S1		S1	S2
Temperature (°C)	50		50	50
Flow rate (mℓ/min)	0.6—0.8		0.3—0.7	0.3—0.7
Column				
Length (cm)	24		25	25
Diameter (mm)	4.6		4.6	4.6
Material	na		na	na
Detection	D1		D2	D2
Reference	1		2	2
Compound	**t_r (min)**	**α[a]**	**t_r (min)**	**t_r (min)**
Lupeol	19.7	2.52	—	—
β-Amyrin	23.4	3.00	—	—
Cycloartenol	25.3	3.27	—	—
α-Amyrin	25.3	3.27	—	—
Germanicol	23.7	3.04	—	—
Cyclolaudenol	27.5	3.53	—	—
Taraxerol	23.2	2.97	—	—
Taraxasterol	14.8	1.90	—	—
Pseudotaraxasterol	18.3	2.42	—	—
24-Methylene-cycloartanol	26.9	3.46	—	—
Bauerenol	26.7	3.43	—	—
Lanosterol	23.4	3.0	—	—
Uvaol	7.8	1.00	—	—
Lupenone	—	—	13.2	—
β-Amyrone	—	—	15.0	—
Cycloartenol	—	—	16.1	—
α-Amyrone	—	—	16.1	—
Friedelin	—	—	18.4	—
Ergosterol	—	—	39.2	—
Cholesterol	—	—	46.9	—
Stigmasterol	—	—	50.0	—
Campesterol	—	—	51.5	—
β-Sitosterol	—	—	56.9	—
β-Glycyrrhetinic acid	—	—	—	3.2
Oleanolic and ursolic acid	—	—	—	4.7
Betulinic acid	—	—	—	5.0

[a] α = ratio t_r/t_r internal standard uvaol.

Column packing	P1	= Zorbax ODS (Dupont)
Solvent	S1	= Methanol containing 0.1% phosphoric acid
	S2	= Methanol-water (95:5) containing 0.1% phosphoric acid
Detection	D1	= UV at 215 nm
	D2	= UV at 215 and 254 nm

REFERENCES

1. **Niemann, G. J. and Baas, W. J.,** *J. Chromatogr. Sci.*, 16, 260, 1978.
2. **Baas, W. J. and Niemann, G. J.,** *J. High Resolut. Chromatogr. Commun.*, 1, 18, 1978.

Table LC 5
STERYL ACETATES

Column packing	P1
Solvent	S1
Temperature	rt
Flow rate (mℓ/min)	1.5
Column	
Length (cm)	30
Diameter (mm, I.D.)	4
Material	SS
Detection	D1

Compound	Grouping	k′
Cholesteryl acetate	$C_{27}\Delta5$	12.7
5-β-Cholestan-3β-yl acetate	$C_{27}\Delta0$ 5β-H	12.2
5-α-Cholestan-3β-yl acetate	$C_{27}\Delta0$ 5-α-H	14.5
5-α-Cholest-7-en-3β-yl acetate	$C_{27}\Delta7$	12.5
Desmosteryl acetate	$C_{27}\Delta5,24$	10.0
Cholesta-5,7-dien-3β-yl acetate	$C_{27}\Delta5,7$	10.6
22-*cis*-Cholesta-5,22-dien-3β-yl acetate	$C_{27}\Delta5,22$	9.2
22-*trans*-Cholesta-5,22-dien-3β-yl acetate	$C_{27}\Delta5,22$	10.2
Campesteryl acetate	$C_{28}\Delta5$	14.4
Brassicasteryl acetate	$C_{28}\Delta5,22$	12.0
24-Methylenecholesteryl acetate	$C_{28}\Delta5,24(28)$	10.9
24-Methylcholesta-5,25-dien-3β-yl acetate	$C_{28}\Delta5,25$	10.5
24-Methylcholesta-5,24-dien-3β-yl acetate	$C_{28}\Delta5,24$	11.4
Ergosteryl acetate	$C_{28}\Delta5,7,22$	10.4
Sitosteryl acetate	$C_{29}\Delta5$	16.0
5α-Stigmast-7-en-3β-yl acetate	$C_{29}\Delta7$	15.3
Stigmasteryl acetate	$C_{29}\Delta5,22$ (24S)	14.4
Poriferasteryl acetate	$C_{29}\Delta5,22$ (24R)	14.4
Fucosteryl acetate	$C_{29}\Delta5,24(28)$ (24E)	13.6
28-Isofucosteryl acetate	$C_{29}\Delta5,24(28)$ (24Z)	13.6
(24Z)-5α-Stigmasta-7,24(28)-dien-3β-yl acetate	$C_{29}\Delta7,24(28)$ (24Z)	13.0
24-Ethylcholesta-5,25-dien-3β-yl acetate	$C_{29}\Delta5,25$	13.0
24-Ethylcholesta-5,22,25-dien-3β-yl acetate	$C_{29}\Delta5,22,25$	11.4

Column packing P1 = Bondapak C_{18}/Corasil (octadecylsilica, 37—50 μm)
Solvent S1 = Methanol-chloroform-water (71:16:13)
Detection D1 = Refractive index detector and UV monitor (254 nm) were coupled in series

REFERENCE

Rees, H. H., Donnahey, P. L., and Goodwin, T. W., *J. Chromatogr.*, 116, 281, 1976.

Table LC 6
ECDYSONE TRITERPENES

Column packing	P1
Solvent	S1
Temperature (°C)	20
Flow rate (mℓ/min)	1
Column	
Length (cm)	160
Diameter (mm)	9
Material	Glass
Detection	D1

Compound	t_r (min)
Inumakilactone	350
Nagilactone C	360
Ecdysterone	420
Inokosterone	455
Makisterone	490
Cyasterone	505
Pterosterone	535
Ecdysone	540
Ponasterone C	550
Ponasterone A	645
Ponasterone B	665

Column packing P1 = Amberlite XAD-2 (20-400 mesh)
Solvent S1 = 200 mℓ ethanol-water (1:4) then 700 mℓ linear gradient of ethanol-water from (1:4 to 7:3)
Detection UV at 230, 250, and 300 nm

REFERENCE

Hori, M., *Steroids,* 14, 33, 1969.

Table LC 7
PENTACYCLIC TRITERPENES

Column packing	P1	P1	P1
Solvent	S1	S2	S3
Temperature	rt	rt	rt
Flow rate (mℓ/min)	2.5	2.5	2.5
Column			
Length (cm)	25	25	25
Diameter (mm)	4.6	4.6	4.6
Material	SS	SS	SS
Detection	UV	UV	UV

Compound		**t_r (min)**	
Acetyl methyl epithelanthate	3.4	—	—
Acetyl methyl-β-glycyrrhetinate	4.7	—	—
Acetyl methyl machaerinate	7.0	—	—
Acetyl methyl betulinate	11.6	—	—
Acetyl methyl oleanolate	11.8	—	—
Acetyl methyl ursolate	11.8	—	—
Benzoyl methyl betulinate	—	4.2	9.2
Benzoyl methyl oleanolate	—	4.2	11.2
α-Amyrin benzoate	—	14.0	21.2
β-Amyrin benzoate	—	14.0	21.2

Column packing P1 = Partisil-10 ODS (Whatman) 10 μm
Solvent S1 = Methanol-water (7:3)
S2 = Methanol-water (4:1)
S3 = Methanol-water (3:1) + 1% silver nitrate
Detection Variable wavelength UV detection

REFERENCE

West, L. G., Templeton, K., and McLaughlin, J. L., *Planta Med.*, 33, 371, 1978.

Table LC 8
TRITERPENE (DIGITALIS) GLYCOSIDES

Column packing	P1	P2	P2	P2
Solvent	S1	S2	S3	S4
Temperature	na	na	na	na
Flow rate (mℓ/min)	1.2	1.4	1.4	1.4
Column				
Length (cm)	15	30	30	30
Diameter (mm, I.D.)	3	3.5	3.5	3.5
Material	na	na	na	na
Detection	D1	D2	D2	D2
Compound		**K′**		
Digitoxigenin	0.35	7.15	2.55	3.77
Gitoxigenin	0.91	2.65	1.10	2.20
Digoxigenin	1.56	1.30	0.30	1.14
Diginatigenin	3.9	0.85	0.17	0.63
Gitaloxigenin	0.35	4.60	1.70	2.64
Digitoxin	0.76	13.80	11.10	7.77
Gitoxin	1.55	4.90	4.56	4.21
Digoxin	1.90	1.95	1.43	1.65
Diginatin	3.40	1.10	0.80	1.10
Gitaloxin	0.86	8.40	7.20	8.00
Lanatoside A	5.50	8.60	9.00	9.52
Lanatoside B	8.71	3.10	3.85	4.89
Lanatoside C	11.27	1.30	1.18	1.99
Lanatoside D	13.40	0.80	0.70	1.16
Lanatoside E	8.60	3.10	3.70	4.54
Desacetyllanatoside C	15.40	0.90	1.40	1.41

Column packing P1 = LiChrosorb Si 100, 5 μm
P2 = Octadecylsilysilica, 10 μm, (Nucleosil C_{18})

Solvent S1 = Methylene chloride-methanol-water (980:80:12)
S2 = Acetonitrile-water (37:63)
S3 = Dioxane-water (45:55)
S4 = Tetrahydrofuran-dioxane-water (44.6:22.3:33)

Detection D1 = UV at 230 nm
D2 = UV at 220 nm

REFERENCE

Erni, F. and Frei, R. W., *J. Chromatogr.*, 130, 169, 1977.

Table LC 9
TRITERPENE (DIGITALIS) GLYCOSIDES-4-NITROBENZOATE DERIVATES

Column packing	P1	P1	P1	P1	P1	P1	P1
Solvent	S1	S2	S3	S4	S5	S6	S7
Temperature	T1	T1	T1	T1	T1	T1	T1
Flow rate	F1	F1	F1	F1	F2	F2	F2
Column							
Length (cm)	20	20	20	20	15	15	15
Diameter (mm, I.D.)	3	3	3	3	3	3	3
Material	na	na	na	na	na	na	na
Detection	D1	D1	D1	D1	D2	D2	D2
Compound[a]	K'						
Digitoxigenin	4.01	2.68	1.93	2.42	4.68	3.87	3.07
Gitoxigenin	5.63	3.08	1.95	3.19	3.55	3.09	2.72
Digoxigenin	6.79	4.19	2.74	3.48	7.41	5.66	4.77
Diginatigenin	—	—	—	—	4.42	3.73	3.52
Gitaloxigenin	—	—	—	—	8.06	6.19	5.01
Digitoxigenin monodigitoxoside	8.04	3.90	2.41	3.89	6.55	5.19	4.31
Digitoxigenin bisdigitoxoside	14.4	5.77	2.99	5.83	9.17	6.92	5.93
Digoxigenin monodigitoxoside	—	—	—	—	9.36	6.96	6.40
Digoxigenin bisdigitoxoside	—	—	—	—	13.0	9.08	8.42
Acetyldigitoxin	18.1	6.59	3.26	6.73	11.6	8.13	6.93
Digitoxin	25.01	8.50	3.85	8.57	12.9	9.14	8.15
Gitoxin	—	—	—	—	10.6	7.75	7.51
Acetyldigoxin	>25	9.52	4.44	9.38	16.5	10.7	9.90
Digoxin	>25	12.0	5.16	11.9	18.3	11.9	11.4
Diginatin	—	—	—	—	2.4	8.77	9.16
Gitaloxin	—	—	—	—	21.1	13.9	12.9
Lanatoside A	>25	23.3	8.17	23.9	32.7	19.1	18.4
Lanatoside B	>25	>25	8.61	>25	27.6	16.0	16.9
Lanatoside C	>25	>25	11.4	>25	46.0	24.6	25.6
Lanatoside D	—	—	—	—	32.1	18.0	20.1
Desacetyl lanatoside A	—	—	—	—	39.5	22.8	23.1
Desacetyl lanatoside B	—	—	—	—	32.3	19.3	21.0
Desacetyl lanatoside C	>25	>25	>20	>25	55.5	28.9	32.2

[a] 4-nitrobenzoates.

Column packing P1 = Merkosorb S1 60 silica gel, 5 μm

Solvent
- S1 = *n*-Hexane-chloroform-methanol (10:1:0.5)
- S2 = *n*-Hexane-chloroform-methanol (10:2:0.5)
- S3 = *n*-Hexane-chloroform-methanol (10:3:0.5)
- S4 = *n*-Hexane-chloroform-methanol (10:1:1)
- S5 = *n*-Hexane-methylene chloride-acetonitrile (10:3.5:2.5)
- S6 = *n*-Hexane-methylene chloride-acetonitrile (10:3:3)
- S7 = *n*-Hexane-chloroform-acetonitrile (30:10:9)

Flow rate
- F1 = 1.5-2.0 mℓ/min
- F2 = 1.35—1.4 mℓ/min

Temperature T1 = 20-23°C

Detection
- D1 = UV at 260 n
- D2 = UV at 254 nm

REFERENCE

Nachtmann, F., Spitzy, H., and Frei, R. W., *J. Chromatogr.*, 122, 293, 1976.

Table LC 10
POLYPRENOLS

Column packing	P1
Solvent	S1
Temperature	na
Flow rate (mℓ/min)	1
Column	
Length (cm)	25
Diameter (cm)	0.4
Material	SS
Detection	D1

Compound	t_r (min)
C_{70} Polyprenol	4.0
C_{75} Polyprenol	4.6
C_{80} Polyprenol	5.3
C_{85} Polyprenol	6.8
C_{90} Polyprenol	8.0
C_{95} Polyprenol	9.4
C_{100} Polyprenol	11.4
C_{105} Polyprenol	13.9

Column packing P1 = μBondapak C_{18} (Waters)
Solvent S1 = Acetone-water (96:4)
Detection D1 = Refractive index

REFERENCE

Sasak, W., Mankowski, T., Chojnacki, T., and Daniewski, W. M., *FEBS Lett.*, 64, 55, 1976.

Table LC 11
CAROTENOIDS

Column packing	P1	P1	P1	P1	P1	P1
Solvent	S1	S2	S3	S4	S5	S6
Temperature	rt	rt	rt	rt	rt	rt
Flow rate (mℓ/min)	1.25	1.25	1.25	1.25	1.25	1.25
Column						
Length (cm)	25	25	25	25	25	25
Diameter (mm)	4.6	4.6	4.6	4.6	4.6	4.6
Material	na	na	na	na	na	na
Detection	D1	D1	D1	D1	D1	D1
Compound			**t_r(min)**			
β,ϵ-Carotene	4.2	—	—	—	—	—
β,β-Carotene	4.2	—	—	—	—	—
β,ψ-Carotene	4.6	—	—	—	—	—
Lycopene	5.0	—	—	—	—	—
2,2′-Dihydroxy β,β-carotene	—	8.0	—	—	—	—
Lutein	—	9.5	—	—	—	—
Zeaxanthin	—	9.7	—	—	—	—
Neo B Lutein	—	—	12.4	—	—	—
Neo A Lutein	—	—	13.1	—	—	—
All-*trans* Lutein	—	—	14.0	—	—	—
Neo U Lutein	—	—	14.8	—	—	—
Neo V Lutein	—	—	15.4	—	—	—
Neo A Bacterioruberin	—	—	—	14.1	—	—
All-*trans* Bacterioruberin	—	—	—	14.6	—	—
Neo U Bacterioruberin	—	—	—	15.1	—	—
Neo V Bacterioruberin	—	—	—	16.0	—	—
Neo W Bacterioruberin	—	—	—	16.3	—	—
Lutein 3′ α-ether	—	—	—	—	17.5	—
Lutein 3′ β-ether	—	—	—	—	17.9	—
8-Auroxanthin	—	—	—	—	—	25.5
8′-Auroxanthin	—	—	—	—	—	25.9
8′-α-Neochrome	—	—	17.5	—	—	—
8′-β-Neochrome	—	—	17.8	—	—	—
Mono-*cis*-neochrome	—	—	18.4	—	—	—
Zeaxanthins	—	—	—	—	—	13.6

Column packing P1 = Silica gel 5-10 μm; best performance Spherisorb 5 μm

Solvent S1 = Hexane containing 0.1% methanol

S2 = Gradient 0-30% acetone in hexane containing 0.1% methanol; rate 10%/min

S3 = Gradient 0-40% acetone in hexane containing 0.1% methanol; rate 3%/min

S4 = Gradient 20-60% acetone in hexane containing 0.1% methanol; rate 1%/min

S5 = Gradient 0-30% acetone in hexane containing 0.1% methanol; rate 1%/min

S6 = Gradient 0-40% acetone in hexane containing 0.1% methanol; rate 1%/min set monochromatically

Detection D1 = Visible spectrophotometry at 400-490 nm depending upon wavelength of maximum absorption for the sample

REFERENCE

Fiksdahl, A., Mortenson, J. T., and Liaaen-Jensen, S., *J. Chromatogr.*, 157, 111, 1978.

Table LC 12
CAROTENOIDS (XANTHOPHYLLS)

Column packing	P1	P2
Solvent	S1	S2
Temperature	na	na
Flow rate (mℓ/min)	1	1
Column		
Length (cm)	30	24
Diameter (mm)	1.8	4.6
Material	SS	SS
Detection	D1	D1

Compound	t_r (min)	
β-Carotene	1.6	4
Lycopene	—	5.3
Chlorin type a	2.5	5.8
Torulene	—	5.8
Spheroidenone	—	18.0
Tetrahydrospirilloxanthin	—	18.5
Spirilloxanthin	—	19.5
Okenone	—	21.0
3,4-Dehydrohodopin	—	23.0
Echinenone	12.5	19.3
Canthaxanthin	15.4	23.4
Chlorin type b	15.8	24.3
Lutein	21.2	29.0
Zeaxanthin	21.2	29.5
Violaxanthin	22.1	30.5
Neoxanthin	—	31.5

Column packing P1 = Silica gel (spherical 10 μm, Spherisorb)
P2 = Silica gel (irregular 5 μm, Partisil)
Solvents S1 = Gradient 2-50% acetone in hexane (concave) over 20 min
S2 = Gradient 1-75% acetone in hexane (concave) over 30 min
Detection D1 = UV at 451 nm

REFERENCE

Hajibrahim, S. K., Tibbetts, P. J. C., Watts, C. D., Maxwell, J. R., Eglinton, G., Colin, H., and Guiochon, G., *Anal. Chem.*, 50, 549, 1978.

Table LC 13
CAROTENOIDS

Column packing	P1	P1
Solvent	S1	S2
Temperature (°C)	28	18
Flow rate (mℓ/min)	2.5	4.0
Column		
Length (cm)	61	61
Diameter (mm, I.D.)	7	7
Material	SS	SS
Detection	D1	D1
Compound	**t_r (min)**	
Fucoxanthin	79.5	—
Neofucoxanthin (A,B)	93.2	—
Diadinoxanthin	135.2	—
Diatoxanthin	157.2	—
Carotenes	265	96
Pheophytin	274.5	92
Violaxanthin	—	30
Lutein	—	53
Neoxanthin	—	23
Chlorophyll a	192.5	72
Chlorophyll b	—	41
Chlorophyll c	58.5	—

Column packing P1 = μBondapak C_{18}-Porasil B (37-75 μm)

Solvent S1 = Methanol-water gradient, 0-20 min. 80% methanol, 21-65 min. 90% methanol, 66-110 min. 95% methanol, 111-170 min. 97.5% methanol, 171-215 min. methanol, 216-250 min. ether-methanol (10:90), 251-270 min. ether-methanol (1:1), 271 min. ether-methanol (75:25)

S2 = Gradient 0-78 min. methanol-water (98:2), 79 min. methanol-ethyl acetate (1:1)

Detection D1 = UV at 440 nm

REFERENCE

Eskins, K., Scholfield, C. R., and Dutton, H. J., *J. Chromatogr.*, 135, 217, 1977.

Table LC 14
CAROTENOIDS

Column packing	P1	P1	P1	P1	P1	P1
Solvent	S1	S1	S1	S1	S1	S1
Temperature	na	na	na	na	na	na
Flow rate	na	na	na	na	na	na
Column						
Length (cm)	20	25	30	35	20	20
Diameter (cm, I.D.)	1	1	1	1	1	1
Material	Glass	Glass	Glass	Glass	Glass	Glass
Detection	D1	D1	D1	D1	D1	D1
Compound			**Relative V_R[a]**			
Echibenone	0.18	0.17	0.18	0.16	0.34	0.34[b]
Canthaxanthin	0.34	0.34	0.35	0.35	0.70	0.71[b]
Astacene	0.46	0.46	0.44	0.44	—	—
Lutein	0.50	0.51	0.50	0.48	1.00	1.00[b]
Violaxanthin	0.67	0.67	0.67	0.68	—	—
Neoxanthin	1.00	1.00	1.00	1.00	—	—

[a] Neoxanthin = 1 for first 4 columns, Lutein = 1 for last two.
[b] Values obtained in the presence of total algal lipid extracts.

Column packing P1 = Silica gel G-Celite 545 (1:1 w/w)
Solvent S1 = Concave gradient, increasing concentration of ethyl acetate-methanol (5:1) in benzene
Detection D1 = Visible at 460 nm

REFERENCE

Davies, B. H., in *Chemistry and Biochemistry of Plant Pigments,* Vol. 2, 2nd ed., Goodwin, T. W., Ed., Academic Press, London, 1976.

Table LC 15
RETINOIDS

Column packing	P1	P2	P3	P4	P5
Solvent	S1	S2	S3	S4	S5
Temperature	18-20°C	18-20°C	rt	rt	rt
Flow rate (mℓ/min)	1	na	0.5	0.1	1.2
Column					
Length (cm)	60	60	15	15	25
Diameter (cm)	2.5	2.5	0.2	0.32	0.2
Material	Glass	Glass	na	SS	SS
Detection	D1	D2	D2	D3	D3
Reference	1	1	2	3	4
Compound	**Ve (mℓ)**		**K′**		**t_r (min)**
Retinol	154	194	5.0	0.97	4.7
α + β-Carotene	85	172	0	—	—
all-*trans*-Retinoic acid	—	—	5.2	—	—
all-*trans*-Retinal	—	—	0.75	—	—
Retinyl acetate,	98	197	0	—	7.6
Retinyl propionate	—	—	—	1.89	—
Retinyl laurate	—	—	—	7.17	—
Retinyl myristate	—	—	—	10.13	—
Retinyl linoleate	—	—	—	11.11	
Retinyl palmitate	88	167	0	14.28	26
Retinyl stearate	—	—	—	20.27	—
N-Acetyl retinyl amine	—	—	—	—	6.0
Retinal acetylhydrazone	—	—	—	—	6.0
Axerophthene (vitamin A hydrocarbon)	—	—	—	—	11.8

Column packing
- P1 = Sephadex LH20
- P2 = S-832 gel
- P3 = MicroPak Si-10 (microparticulate silica)
- P4 = RSIL C_{18}HL 10 μm (octadecylsilica with 18% bonded organic material)
- P5 = Spherisorb 5 μm (octadecylsilica)

Solvent
- S1 = Chloroform
- S2 = Tetrahydrofuran
- S3 = Petroleum ether-dichloromethane-isopropanol (80:19.3:0.7)
- S4 = Methanol
- S5 = Acetonitrile-water, (79:21) until 13 min. then (98:2)

Detection
- D1 = UV, VIS, IR
- D2 = UV (254 nm)
- D3 = UV (325 nm)

REFERENCES

1. **Holasova, M. and Blattna, J.,** *J. Chromatogr.*, 123, 225, 1976.
2. **DeRuyter, M. G. M. and De Leenheer, A. P.,** *Clin. Chem. (Winston-Salem, N.C.)*, 22, 1953, 1976.
3. **DeRuyter, M. G. M. and De Leenheer, A. P.,** *Clin. Chem. (Winston-Salen, N.C.)*, 24, 1920, 1978.
4. **Roberts, A. B., Nichols, M. D., Frolik, C. A., Newton, D. L., and Sporn, M. B.,** *Cancer Res.*, 38, 3327, 1978.

Table LC 16
ISOMERIC RETINOIDS

Column packing	P1	P1	P1	P1	P1
Solvent	S1	S2	S3	S4	S5
Temperature	rt	na	na	na	na
Flow rate (mℓ/min)	2	na	na	na	na
Column					
Length (cm)	25	25	25	25	25
Diameter (cm)	0.79	0.79	0.21	0.79	0.79
Material	na	na	na	na	na
Detection	D1	D1	D1	D2	D3
Reference	1	2	2	2	2

Compound	r_a	r_b	r_c	r_d	r_e	r_f
13-*cis*-Retinal	2.19	2.37	0.70	—	—	—
11-*cis*-Retinal	2.50	2.70	0.77	—	—	—
9-*cis*-Retinal	2.75	2.98	0.80	—	—	—
all-*trans*-Retinal	3.53	3.82	1.00	—	—	—
11-*cis*-Retinol	—	—	1.18	2.10	2.02	1.72
13-*cis*-Retinol	—	—	1.18	2.10	2.09	1.78
9-*cis*-Retinol	—	—	1.52	2.75	2.71	2.34
all-*trans*-Retinol	—	—	1.59	2.92	2.85	2.47

Note: Retention times (r = 1.0) relative to: a = tetraphenylethylene (5.4 min); b = 2,6-ditert-butyl-*p*-cresol; c = all *trans*-retinal (14.1 min); d = *o*-xylenol (2.4 min); e = α-naphthol (7.8 min); f = *o*-vanillin (8.3 min).

Column packing	P1 =	Zorbax SIL
Solvent	S1 =	12% diethyl ether-hexane (12:88)
	S2 =	Diethyl ether-*n*-hexane (1:4)
	S3 =	Ethyl acetate-*n*-hexane (12:88)
	S4 =	Ethyl acetate-methylene chloride-*n*-hexane (7.5:9.3:82.2)
	S5 =	Ethyl acetate-methylene chloride-*n*-hexane (6.2:7.7:86.1)
Detection	D1 =	UV at 254 or/and 380 nm
	D2 =	UV at 328 nm
	D3 =	UV at 320 nm

REFERENCES

1. **Tsukida, K., Kodama, A., and Ito, M.,** *J. Chromatogr.*, 134, 331, 1977.
2. **Tsukida, K., Kodama, A., Ito, M., Kawamoto, M., and Takahashi, K.,** *J. Nutr. Sci. Vitaminol.*, 23, 263, 1977.

Table LC 17
ISOMERIC RETINOIDS

Column packing	P1	P1	P1	P1	P1	P1	P1
Solvent	S1	S2	S3	S4	S5	S6	S7
Temperature	na	na	na	na	na	na	na
Flow rate	na	na	na	na	na	na	na
Column							
Length (cm)	25	25	25	25	25	25	25
Diameter (cm)	3	3	3	3	3	3	3
Material	SS	SS	SS	SS	SS	SS	SS
Detection	D1	D1	D1	D1	D1	D1	D1
Compound				**Ki[a]**			
13-*cis*-Retinal	2.95	2.17	1.35	—	—	—	—
11-*cis*-Retinal	3.17	2.40	1.50	—	—	—	—
9-*cis*-Retinal	3.96	2.84	1.69	—	—	—	—
all-*trans*-Retinal	5.11	3.59	2.11	—	—	—	—
13-*cis*-Retinol	—	21.01	9.96	4.37	—	—	—
11-*cis*-Retinol	—	20.14	9.52	4.17	—	—	—
9-*cis*-Retinol	—	23.95	11.38	4.85	—	—	—
all-*trans*-Retinal	—	28.07	13.29	5.62	—	—	—
13-*cis*-Retinyl palmitate	—	—	—	—	8.81	2.70	2.28
11-*cis*-Retinyl palmitate	—	—	—	—	8.81	2.84	2.41
9-*cis*-Retinyl palmitate	—	—	—	—	11.15	3.33	2.79
all-*trans*-Retinyl palmitate	—	—	—	—	13.24	4.08	3.36

[a] Ki = ratio of the retention time of the component over the retention time of benzene minus 1.

Column packing	P1 =	Si 60 silica gel (5 μm)
Solvent	S1 =	Dioxane-*n*-hexane (1.25:98.75)
	S2 =	Dioxane-*n*-hexane (2.5:97.5)
	S3 =	Dioxane-*n*-hexane (95:5)
	S4 =	Dioxane-*n*-hexane (9:1)
	S5 =	*n*-Hexane
	S6 =	Dioxane-*n*-hexane (0.05:99.95)
	S7 =	Dioxane-*n*-hexane (0.01:99.9)
Detection	D1 =	UV retinyl palmitate and retinol (320 nm); retinal (360 nm)

REFERENCE

Paanakker, J. E. and Groenendijk, G. W., *J. Chromatogr.*, 168, 125, 1979.

Table LC 18
TOCOPHEROLS

Column packing	P1	P2	P3	P4
Solvent	S1	S2	S3	S4
Temperature	na	40°C	rt	rt
Flow rate (mℓ/min)	2.39	2	0.66	2.5
Column				
Length (cm)	25	25	50	30
Diameter (mm)	4.6	4	2.2	3.9
Material	SS	na	SS	SS
Detection	D1	D2	D3	D4
Reference	1	2	3	4

Compound	r	k′	t_r (min)	t_r (min)
α-Tocopherol	1.5 (6 min)	3.54	10.6	5.6
β-Tocopherol	2.1	3.01	13.2	5.0
γ-Tocopherol	2.3	3.01	14.8	5.0
δ-Tocopherol	4.3	—	18.0	3.0
Tocol	—	2.10	—	7.4
α-Tocopheryl quinone	—	—	—	3.0
α-Tocopheryl acetate	—	—	—	7.4
all-*trans*-Retinol	—	1.06	—	2.4
Retinyl acetate	—	—	—	3.0

Column packing
- P1 = Partisil PXS10 10 μm
- P2 = R Sil C_{18} 10 μm
- P3 = Micro Pak Si 10 μm
- P4 = μBondapak C_{18}

Solvent
- S1 = Hexane-methanol (97:3)
- S2 = Methanol
- S3 = Hexane-diisopropyl ether linear gradient, initial solvent ratio (95:5), slope 1% diisopropyl ether/min
- S4 = H_2O-methanol (95:5)

Detection
- D1 = Fluorescence excitation 295 nm, emission 340 nm
- D2 = UV at 292 nm
- D3 = UV at 297 nm
- D4 = UV at 280 nm

REFERENCES

1. **Vatassery, G. T., Maynard, V. R., and Hagen, D. F.,** *J. Chromatogr.*, 161, 299, 1978.
2. **De Leenheer, A. P., De Bevere, V. O. R. C., De Ruyter, M. G. M., and Claeys, A. E.,** *J. Chromatogr.*, 162, 408, 1979.
3. **Matsuo, M. and Tahara, Y.,** *Chem. Pharm. Bull.*, 25, 3381, 1977.
4. **Bieri, J. G., Tolliver, T. J., and Catignani, G. L.,** *Am. J. Clin. Nutr.*, 32, 2143, 1979.

Table LC 19
TOCOPHEROLS

Column packing	P1	P2	P3
Solvent	S1	S2	S3
Temperature	rt	18-20°C	rt
Flow rate (mℓ/min)	1.5	1	1
Column			
Length (cm)	150	60	25
Diameter (cm, I.D.)	0.02	2.5	0.21
Material	SS	Glass	SS
Detection	D1	D2	D3
Reference	1	382	3

Compound	Retention Ratio[a]		Ve (mℓ)
α-Tocopherol	1.00	118	0.69
β-Tocopherol	1.54	—	—
γ-Tocopherol	1.88	—	—
δ-Tocopherol	3.00	—	—
α-Tocopheryl acetate	—	91	1.00
δ-Tocopheryl acetate	—	—	0.74
γ + β-Tocopheryl acetate	—	—	0.83

[a] Retention time of α-tocopherol is 10 min.

Column Packing P1 = Corasil II
P2 = Sephadex LH-20
P3 = Octadecyl-silica

Solvent S1 = Diisopropyl ether-*n*-hexane (5:95)
S2 = Chloroform
S3 = H_2O-methanol (3:97)

Detection D1 = Fluorimetry, primary interference filter at 295 nm, secondary filter 340 nm
D2 = UV, VIS, and IR
D3 = UV

REFERENCES

1. **Van Niekerk, P. J.,** *Anal. Biochem.*, 52, 533, 1973.
2. **Holasova, M. and Blattna, J.,** *J. Chromatogr.*, 123, 225, 1976.
3. **Eriksson, T. and Sörensen, B.,** *Acta Pharm. Suec.*, 14, 475, 1977.

Table LC 20
VITAMIN K AND DERIVATIVES

Column packing	P1	P1
Solvent	S1	S2
Temperature	na	na
Flow rate	na	na
Column		
Length (cm)	30	20
Diameter (cm)	3	2
Material	na	na
Detection	D1	D1

Compound	**t_r (min)**	
Menaquinone-4	27.5	—
2,3-Epoxymenaquinone-4	20.0	—
Menaquinone-3	17.0	—
2,3-Epoxymenaquinone-3	12.0	—
Phylloquinone	—	9.6
2,3-Epoxymenaquinone	—	7.2
Menadione	5.0	—

Column packing P1 = μBondapak C_{18} 10 μm
Solvent S1 = Acetonitrile-water (85:15)
S2 = Methanol
Detection D1 = UV at 254 nm

REFERENCE

Donnahey, P. L., Burt, V. T., Rees, H. H., and Pennock, J. F., *J. Chromatogr.*, 170, 272, 1979.

Table TLC 1
MONOTERPENE HYDROCARBONS

Layer	L1	L2
Solvent	S1	S2
Detection	D1	D2
Compound	**$R_F \times 100$**	
p-χ-Dimethylstyrene	33	74
p-Cymene	39	80
Myrcene	42	50
Sabinene	43	47
Limonene	44	67
Terpinolene	46	64
γ-Terpinene	48	70
α-Phellandrene	49	66
Verbenene	50	72
β-Pinene	51	69
Camphene	52	71
Δ^3-Carene	53	76
α-Pinene	55	80

Layer L1 = 15% (w/w) thallous nitrate on silica gel G
L2 = 7.8% (w/w) silver nitrate on silica gel G
Solvent S1 = Hexane
S2 = Benzene
Detection D1 = Iodine
D2 = Antimony pentachloride in chloroform

REFERENCE

Baines, D. A. and Jones, R. A., *J. Chromatogr.*, 47, 130, 1970.

Table TLC 2
ISOPRENOID ALCOHOLS AND ALDEHYDES

Layer	L1	L1
Solvent	S1	S2
Detection	D1,D2	D3
Compound	**$R_F \times 100$**	
Isopentenol	44—45	—
Dimethylallyl alcohol	71—72	—
Dimethylvinylcarbinol	52—54	—
Nerol	70—72	—
Geraniol	61—62	—
Linalool	68—69	—
2-*cis*, 6-*trans*-Farnesol	57—58	—
2,6-*trans,trans*-Farnesol	46—47	—
trans-Nerolidol	82—83	—
Neral, geranial	—	41—43
Farnesals	—	31—33
Dimethylallyl aldehyde	—	36—38

Layer L1 = Silica gel G on GF 257
Solvent S1 = Ethyl acetate
S2 = Benzene
Detection D1 = UV (260 nm) on GF 254
D2 = 10% antimony pentachloride in chloroform
D3 = 2,4-dinitrophenylhydrazine in ethanol
Technique Ascending 15 cm

REFERENCE

Cardemil, E., Vicuna, J. R., Jabalquinto, A. M., and Cori, O., *Anal. Biochem.*, 59, 636, 1974.

Table TLC 3
OXYGENATED MONOTERPENES

Layer	L1	L1	L1	L1
Solvent	S1	S2	S3	S4
Detection	D1-D4	D1	D1	D1
Reference	1,2	2	2	3
Compound	**R_F × 100**			
Menthene	63	—	—	—
Dihydrocarvone	48	—	—	—
Pulegone	40	—	—	—
Carvone	38	—	—	—
Piperitone	30	—	—	—
Carvenone	28	—	—	—
Piperitenone	20	—	—	—
(−)Borneol	32	65	58	—
(−)Borneol carbamate	18	68	53	—
(+)Isoborneol	38	61	58	—
(+)Isoborneol carbamate	12	64	53	—
(−)Isopulegol	42	65	56	—
(−)Isopulegol carbamate	23	68	61	—
cis-Linalool oxide	25	62	47	—
cis-Linalool oxide carbamate	16	50	49	—
Cedrol	41	66	58	—
Cedrol carbamate	11	63	57	—
α-Terpineol	28	65	55	29
α-Terpineol carbamate	20	61	57	—
4-Terpineol	32	63	51	—
4-Terpineol carbamate	25	64	53	—
Citronellol	32	64	55	24
Citronellol carbamate	21	—	—	—
Linalool	34	67	62	43
Linalool carbamate	21	—	56	—
Limonene	—	—	—	96

Layer L1 = Silica gel G

Solvent S1 = Benzene

S2 = Ethyl acetate

S3 = Ethyl acetate-benzene (3:1)

S4 = *n*-Hexane-ethyl acetate (85:15)

Detection D1 = 1% Vanillin in conc. H_2SO_4 or 3% vanillin in 0.5% conc. H_2SO_4-absolute methanol

D2 = 20% Antimony (V) chloride in carbon tetrachloride

D3 = 0.5 mℓ Anisaldehyde in acetic acid-methanol-conc. sulfuric acid (10:85:5)

D4 = 1% *p*-Dimethylaminobenzaldehyde in conc. sulfuric acid

REFERENCES

1. **Rothbacher, H. and Suteu, F.,** *J. Chromatogr.*, 100, 236, 1974.
2. **Gueldner, R. C., Hutto, F. Y., Thompson, A. C., and Hedin, P. A.,** *Anal. Chem.*, 45, 376, 1973.
3. **Karawya, M. S., Balbaa, S. I., and Hifnawy, M. S.,** *J. Pharm. Sci.*, 60, 381, 1971.

Table TLC 4
CAMPHOR METABOLITES

Layer	L1	L1	L1
Solvent	S1	S2	S3
Detection	D1,D2	D1,D2	D1,D2
References	1	1	2
Compound		**R_F × 100**	
cis-2-*exo*,3-*exo*-Camphanediol	25	40	—
2-*endo*-Hydroxyepicamphor	35	50	35
Borneol	59	—	57
Isoborneol	—	—	61
Epiborneol	49	—	49
Camphor	70	—	70
cis-Diol acetonide	—	75	—
Norcamphor	—	—	54
Epicamphor	—	—	67
Camphorquinone	—	—	36
Camphane-2,5-dione	—	—	38
Epi-isoborneol	—	—	55
endo-Norborneol	—	—	27
3-*endo*-Hydroxycamphor	—	—	35
5-*endo*-Hydroxycamphor	—	—	13

Layer L1 = Silica gel H
Solvent S1 = Light petroleum (bp 60-80°C)-ethyl acetate (3:1)
S2 = Light petroleum (bp 60-80°C)-ethyl acetate-ethanol (10:2:1)
S3 = *n*-Hexane-ethyl acetate (3:1)
Detection D1 = 10% Alcoholic phosphomolybdic acid, heat at 110°C for 1 min
D2 = Iodine

REFERENCES

1. **Robertson, J. S. and Solomon, E.,** *Biochem. J.*, 121, 503, 1971.
2. **Robertson, J. S. and Hussain, M.,** *Biochem. J.*, 113, 57, 1969.

Table TLC 5
OXYGENATED LIMONENE METABOLITES

Layer	L1	L1	L1	L1	L1	L2
Solvent	S1	S2	S3	S4	S5	S6
Detection	D1,D2	D1,D2	D1,D2	D1,D2	D1,D2	D1,D2
References	1	1	1	1,2	1,2	2
Compound			**$R_F \times 100$**			
d-Limonene	96	90	84	90	97	93
p-Mentha-1,8-dien-10-ol	87	66	71	80	92	—
p-Menth-1-ene-8,9-diol	75	18	16	59	85	—
Perillic acid	81	53	60	80	88	—
Perillic acid-8,9-diol	53	8	6	30	86	54
p-Mentha-1,8-dien-10 yl-β-D-gluco pyranosiduronic acid	26	0	0	0	72	26
Acetylated methyl ester of above	93	49	68	89	93	—
8-hydroxy-*p*-menth-1-en-9-yl-β-D-gluco pyranosiduronic acid	10	0	0	0	53	11
Acetylated methyl ester of above	91	15	32	77	89	—
2-hydroxy-p-menth-8-en-7-oic acid	—	—	—	24[a]	78[a]	77
Perillylglycine	—	—	—	8[a]	76[a]	72
Perillyl-β-D-glucopyranosiduronic acid	—	—	—	0[a]	51[a]	20

[a] Layer L2 used for these determinations.

Layer L1 = Silica gel G
L2 = Silica gel 60 F 254
Solvent S1 = Benzene-ethyl acetate-methanol-formic acid (sp. gr. 1.22) (6:14:1:2)
S2 = *n*-Hexane-ethyl acetate-acetic acid (70:30:0.01)
S3 = Benzene-ethyl acetate-acetic acid (30:15:0.02)
S4 = Benzene-ethyl acetate-methanol-acetic acid (11:1:1:0.1)
S5 = *N*-Hexane-ethyl acetate-methanol-formic acid (sp. gr. 1.22) (2:7:3:1)
S6 = Benzene-ethyl acetate-methanol-formic acid (sp. gr. 1.22) (5:10:1:1.5)
Detection D1 = Anisaldehyde-methanol-conc. H_2SO_4 (1:1:50), heat at 130°C for 5 min
D2 = 1% Potassium permanganate in 6% aqueous sodium carbonate

REFERENCES

1. **Kodama, R., Noda, K., and Ide, H.,** *Xenobiotica,* 4, 85, 1974.
2. **Kodama, R., Yano. T., Furukawa, K., Noda, K., and Ide, H.,** *Xenobiotica,* 6, 377, 1976.

Table TLC 6
IRIDOID MONTERPENES

Layer	L1	L1	L1
Solvent	S1	S2	S3
Compound		**R_F × 100**	
Loganin	29	29	45
Geniposide	33	32	39
Gentiopicroside	38	32	43
Secologanin	44	37	54
7-Ketologanin	48	42	—
Asperuloside	48	34	44
Verbenalin	—	42	50

Layer L1 = Silica gel G
Solvents S1 = Chloroform-methanol (4:1)
S2 = Chloroform-methanol-benzene (8:2:1), double development
S3 = Ethyl acetate-methanol (4:1)

Detection Concentrated sulfuric acid and heat at 100°C for 10 min. Iridoids give characteristic brown, violet and red colors

REFERENCE

Coscia, C. J., in *Chromatography,* Heftmann, E., Ed., 3rd ed., Van Nostrand-Reinhold, New York, 1975, 594.

Table TLC 7
SESQUITERPENE HYDROCARBONS

Layer	L1	L1	L1	L2	L2	L2	L3	L3	L3	L4	L4	L4
Solvent	S1	S2	S3	S1	S2	S3	S1	S2	S3	S1	S2	S3
Compound						$R_F \times 100$						
Longicyclene	75	—	—	68	—	—	77	—	—	76	—	—
Isolongifolene	66	—	—	58	—	—	67	—	—	66	—	—
Longifolene	44	—	—	31	—	—	32	—	—	14	—	—
α-Gurijunene	—	82	—	—	80	—	—	79	—	—	78	—
α-Bergamotene	—	74	—	—	62	—	—	71	—	—	61	—
α-Bisabolene	—	37	—	—	18	—	—	32	—	—	16	—
α-Himachalene	—	—	69	—	—	56	—	—	70	—	—	55
β-Himachalene	—	—	61	—	—	46	—	—	62	—	—	46

Layer L1 = 15% (w/w) Silver nitrate on silica gel-calcium sulfate (9:1)
L2 = 15% (w/w) Silver nitrate on silica gel
L3 = 15% (w/w) Silver perchlorate on silica gel-calcium sulfate (9:1)
L4 = 15% (w/w) Silver perchlorate on silica gel
Solvent S1 = Light petroleum
S2 = Light petroleum-benzene (1:1)
S3 = Light petroleum-benzene (4:1)

Detection Chlorosulfonic acid-acetic acid (1:2), heat 10 min at 130°

REFERENCE

Prasad, R. S., Gupta, A. S., and Dev, S., *J. Chromatogr.*, 92, 450, 1972.

Table TLC 8
OXYGENATED SESQUITERPENES

Layer	L1	L1	L1	L1	L1	L1	L1
Solvent	S1	S2	S3	S4	S5	S6	S7
Detection	D1	D2,D3	D2,D3	D2,D3	D2,D3	D2,D3	D2,D3
References	1	2	2	2	2	2	2
Compound				**R_F × 100**			
Arbusculin-B	69	—	—	—	—	—	—
Arbusculin-C	49	—	—	—	—	—	—
Arbusculin-A	29	—	—	—	—	—	—
Rothin-A	17	—	—	—	—	—	—
Rothin-B	9	—	—	—	—	—	—
Rishitin	—	42	28	25	64	21	66
Rishitin M1	—	27	6	7	45	6	57
Rishitin M2	—	21	5	6	44	6	57

Layer L1 = Silica gel G
Solvent S1 = Light petroleum-chloroform-ethyl acetate (2:2:1)
S2 = Ethyl acetate
S3 = Diethyl ether
S4 = Ethyl acetate-cyclohexane (1:1)
S5 = Diethyl ether-methanol (9:1)
S6 = Chloroform-acetone (85:15)
S7 = Ethyl acetate-methanol (8:2)
Detection D1 = Conc. sulfuric acid and heat 100-115°C for 5 min
D2 = 5% Cerium (IV) sulfate in 1 *M* sulfuric acid
D3 = Conc. sulfuric acid followed by 5% cerium (IV) sulfate in 1 *M* sulfuric acid and heat

REFERENCES

1. **Kelsey, G., Morris, M. S., Bhadane, N. R., and Shafizadeh, F.,** *Phytochemistry,* 12, 1345, 1973.
2. **Ishiguri, Y., Tomiyama, K., Doke, N., Murai, A., Katsui, K., Yagihashi, F., and Masamune, T.,** *Phytopathology,* 68, 720, 1978.

Table TLC 9
SESQUITERPENE LACTONES

Layer	L1	L2	L1	L2	L1	L2
Solvent	S1	S1	S2	S2	S3	S3
Detection	D1-D3	D1-D3	D1-D3	D1-D3	D1-D3	D1-D3
Compound			**$R_F \times 100$**			
Gafrinin	44	38	45	45	37	31
Geigerin	36	31	30	32	30	26
Geigerinin	20	19	15	12	11	06
Vermeerin	26	25	61	53	35	29
Dihydrogriescenin	61	49	68	61	51	43
Griesenin	51	43	66	57	46	41

Layer L1 = Silica gel G
L2 = Silica gel 60F-254
Solvent S1 = Chloroform-methanol (96:4)
S2 = Hexane-isobutanol (7:3)
S3 = 2-Butanone-light petroleum (b.p. 60-80), (1:1)
Detection D1 = UV light at 254 and 360 nm
D2 = Iodine
D3 = Anisaldehyde-sulfuric acid, heat at 110°C for 10 min and UV inspection

REFERENCE

Von Jeney De Boresjeno, N. L. T. R. M., Potgieter, D. J. J., and Vermeulen, N. M. J., *J. Chromatogr.*, 94, 255, 1974.

Table TLC 10
DITERPENE HYDROCARBONS

Layer	L1	L2	L3	L4	L5
Solvent	S1	S2	S2	S2	S2
Detection	D1,D2	D3	D3	D3	D3
References	1	2	2	2	2
Compound	$r_{kaurene}$[a]	$R_F \times 100$			
Kaurane	1.22	—	—	—	—
Trachylobane	1.22	—	—	—	—
Kaurene	1.00	—	—	—	—
Phyllocladene	1.00	—	—	—	—
Pimaradiene	0.96	—	—	—	—
Atiserene	0.91	—	—	—	—
Isopimaradiene	0.88	—	—	—	—
Sandaracopimaradiene	0.83	—	—	—	—
Isophyllocladene	0.69	—	—	—	—
Isokaurene	0.41	—	—	—	—
Beyerene	0.40	—	—	—	—
Isoatiserene	0.31	—	—	—	—
Cembrene	—	85	81	86	83
Cembrene A	—	65	41	42	16

[a] R_F of kaurene varied between 0.65 and 0.75.

Layer L1 = Silical gel G - 3% silver nitrate (w/w)
L2 = 15%(w/w) Silver nitrate on silica gel-calcium sulfate (9:1)
L3 = 15% (w/w) Silver nitrate on silica gel
L4 = 15 % (w/w) Silver perchlorate on silica gel-calcium sulfate (9:1)
L5 = 15 % (w/w) Silver perchlorate on silica gel

Solvent S1 = *n*-Hexane-benzene (7:3)
S2 = Ethyl acetate-benzene (15:85)

Detection D1 = 10% Phosphomolybdic acid in methanol
D2 = Iodine
D3 = Chlorosulfonic acid-acetic acid (1:2), heat 10 min at 130°C

REFERENCES

1. **Robinson, D. R. and West, C. A.,** *Biochemistry*, 9, 70, 1969.
2. **Prasad, R. S., Gupta, A. S., and Dev, S.,** *J. Chromatogr.*, 92, 450, 1972.

Table TLC 11
DITERPENE GIBBERELLINS

Layer	L1	L1	L2	L1	L1	L3	L2
Solvent	S1	S2	S3	S4	S5	S6	S7
Compound				**$R_F \times 100$**			
Gibberellin A_1	9	31	20	25	34	62	0
Gibberellin A_3	9	29	20	29	31	55	0
Gibberellin A_4	44	53	—	41	50	100	54
Gibberellin A_7	44	53	—	41	49	100	45
Isogibberellin A_3	9	32	17	25	29	52	0
Isogibberellin A_7	44	53	—	35	48	100	43
allo-Gibberic acid	54	—	—	35	52	—	—

Layer L1 = Silica gel G
L2 = Silica gel G impregnated with silver nitrate (125 mg/mℓ water)
L3 = Kieselguhr G

Solvent S1 = Isopropyl ether-acetic acid (19:1)
S2 = Isopropyl ether-acetone-acetic acid (90:3:1)
S3 = Ethyl acetate-chloroform-acetic acid (50:10:1)
S4 = *N*-Butanol-1. 5 *N* ammonium hydroxide (1:1, organic phase)
S5 = Benzene-acetic acid-water (50:40:3)
S6 = Benzene-propionic acid-water (8:3:5) (plates were equilibrated 3 hr before irrigation with the organic phase)
S7 = Carbon tetrachloride-acetic acid-water (8:3:5) (plates were equilibrated 18 hr before irrigation with the organic phase)

Detection Sulfuric acid-fluorescence at rt and 110°C for 15 min

REFERENCE

Pitel, D. W., Vining, L. C., and Arsenault, G. P., *Can. J. Biochem.*, 49, 185, 1971.

Table TLC 12
GRAYANOTOXIN DITERPENES

Layer	L1	L1	L1	L1	L2	L2	L2	L2	L3	L3	L3
Solvent	S1	S2	S3	S4	S1	S2	S3	S4	S5	S6	S7
Detection	D1-D3	D1-D3	D1-D3	D1-D3	D1-D3	D1-D3	D1-D3	D1-D3	D1-D3	D1-D3	D1-D3
Compound						$R_F \times 100$					
Grayanotoxin III (GIII)	30	3	48	8	41	3	57	5	2	10	6
Grayanotoxin I (GIII 14-acetate)	41	6	56	22	55	9	66	19	5	24	18
GIII 6-acetate	39	5	56	19	50	4	64	11	4	24	19
GIII 6,14-diacetate	59	18	67	39	66	21	75	39	24	59	56
GIII 3,6,14-triacetate	67	35	73	49	75	39	79	60	66	85	88
GIII 6-propionate	48	6	61	24	57	6	68	14	10	34	30
GIII 3,6-dipropionate	65	17	71	41	68	9	75	24	54	80	82
GIII 6-propionate 14-acetate	65	25	70	46	70	25	77	41	43	72	72
GIII 6-butyrate	53	7	62	25	61	6	73	16	18	53	42
GIII 3,6-dibutyrate	75	37	82	60	76	33	82	65	87	91	93
GIII 6-butyrate 14-acetate	69	27	75	51	73	24	79	51	62	80	81
GIII 6-isobutyrate	54	9	62	27	62	7	75	18	19	53	41
GIII 6-benzoate	86	76	84	77	82	79	88	88	99	99	99
Rhodojaponin I	66	39	70	54	75	45	77	66	65	83	83
Rhodojaponin VI	32	2	49	11	29	1	46	3	2	7	3
Lyoniol A	45	14	59	31	62	17	70	33	12	40	36

Layers
- L1 = Silica gel G
- L2 = Aluminum oxide G
- L3 = Kieselguhr G UV/G_{254} eluted with 20% ethylene glycol in acetone

Solvent
- S1 = Chloroform-isopropanol-water-ether (15:12:1:60)
- S2 = Ether-benzene (2:1), double development
- S3 = Ethyl acetate-isopropanol-water (80:24:6)
- S4 = Hexane-benzene-ether-acetone (1:3:9:2)
- S5 = Toluene-ether (1:2)
- S6 = Ethyl acetate-ether (3:2)
- S7 = Ethyl acetate-cyclohexane (3:1)

Detection D1 = Spray with 60% sulfuric acid. heat at 110°C for 5 min
D2 = Spray with Godin's reagent (1% vanillin in ethanol), oversprayed with 3% ethanolic perchloric acid; heat at 80°C for 4 min
D3 = 1% vanillin in ethanol oversprayed with 60% sulfuric acid; heat at 110°C for 5 min

REFERENCE

Kinghorn, A. D., Jawad, F. H., and Doorenbos, N. J., *J. Chromatogr.*, 147, 299, 1978.

Table TLC 13
DITERPENE ACETATES

Layer	L1	L1	L1	L2	L3	L3	L3	L3	L3	L3	L3
Solvent	S1	S2	S3	S4	S5	S6	S7	S8	S9	S10	S11
Detection	D1-D4	D1-D4	D1-D4	D1-D4	D1-D4	D1-D 4	D1-D4	D1-D4	D1-D4	D1-D4	D1-D4
Compound						$R_F \times 100$					
Phorbol triacetate	16	48	64	49	5	31	53	6	30	22	35
12-Deoxy-phorbol diacetate	15	51	68	50	6	33	53	10	38	29	39
4-Deoxy-4α-phorbol triacetate	21	52	55	53	12	51	72	11	52	36	59
4α-Phorbol triacetate	7	26	44	39	1	4	10	1	3	2	5
4α-Phorbol tetraacetate	26	49	57	55	11	54	76	7	45	43	51
Crotophorbolene monoacetate	8	34	60	41	2	10	20	2	11	7	111
Ingenol triacetate	44	65	72	59	32	69	80	37	70	74	86
Compound A_1	3	31	71	33	2	7	6	1	6	4	9

Layer
- L1 = Silica gel G
- L2 = Silica gel H
- L3 = Alumina E

Solvent
- S1 = Chloroform-ether (95:5)
- S2 = Diethyl ether-ethyl acetate-hexane (1:1:1)
- S3 = Hexane-isopropyl alcohol (2:1)
- S4 = Chloroform-ethyl acetate (2:3)
- S5 = Toluene-ethyl acetate (9:2)
- S6 = Chloroform-acetone-benzene (95:5:50)
- S7 = Chloroform
- S8 = Hexane-ether-benzene (1:2:1), developed three times
- S9 = Benzene-hexane-ether-ethyl acetate (20:40:15:30), developed three times
- S10 = Chloroform-ether-benzene (1:3:3), developed three times
- S11 = Ethyl acetate-benzene (1:3), developed four times

Detection
- D1 = 60% (w/w) sulfuric acid, heated at 110°C for 15 min
- D2 = 5% Vanillin in conc. sulfuric acid, heated at 110°C for 5 min
- D3 = 1% Anisaldehyde in conc. sulfuric acid-acetic acid (2:98), heat at 110°C for 10 min
- D4 = Methanol-conc. sulfuric acid (1:1)

All plates viewed with visible and UV light (at 366 nm)

REFERENCE

Evans, F. J. and Kinghorn, A. D., *J. Chromatogr.*, 87, 443, 1973.

Table TLC 14
PHYTOLS

Layer	L1	L1	L1
Solvent	S1	S2	S3
Detection	D1-D2	D1-D2	D1-D2
Compound		$R_F \times 100$	
Phytol	0	24	16
Isophytol	0	29	21
O-Methylphytol	0	50	40
O-Methylisophytol	0	57	47
Phytanyl chloride	57	76	74
Phytene	77	76	74

Layer	L1	= Silica gel H
Solvent	S1	= Petroleum ether
	S2	= Chloroform
	S3	= Chloroform-benzene (1:1)
Detection	D1	= Water
	D2	= Rhodamine G, viewed with UV light

REFERENCE

Joo, C. N., Park, C. E., Kramer, J. K. G., and Kates, M., *Can. J. Biochem.*, 51, 1527, 1973.

Table TLC 15
PIERCIDINS (DITERPENES)

Layer	L1	L1
Solvent	S1	S2
Detection	D1-D3	D1-D3
Piercidins	**R_F × 100**	
B_4	58	67
B_3	56	64
B_2	56	64
B_1	54	62
D_4	45	57
D_3	37	54
D_2	34	46
D_1	30	40
A_4	54	55
A_3	48	49
A_2	41	43
A_1	34	33
C_4	16	22
C_3	14	18
C_2	12	16
C_1	10	12

Layer L1 = Silica gel GF_{254}
Solvent S1 = Benzene-ethyl acetate (93:7)
S2 = Hexane-tetrahydrofuran (4:1)
Detection D1 = Dragendorff reagent
D2 = 3% Aqueous potassium permanganate
D3 = UV light (254 nm)

REFERENCE

Yoshida, S., Yoneyama, K., Shiraishi, S., Watanabe, A., and Takahashi, N., *Agric. Biol. Chem.*, 41, 849, 854, 1977.

Table TLC 16
ISOPRENYL PHOSPHATES

Layer	L1	L1	L1
Solvent	S1	S2	S3
Detection	D1,D2	D1,D2	D1,D2
Compound		**$R_F \times 100$**	
Phytanylpyrophosphate tripotassium salt[a]	25, 14, 8	40, 22, 13	9
Phytylpyrophosphate triammonium salt	25, 15, 7	39, 26, 14	9
Phytanylmonophosphate dipotassium salt	41	47	26
Phytylmonophosphate diammonium salt	45	47	26

[a] The pyrophosphates give three spots in S1 and S2, the central one predominating.

Layer L1 = Silica gel H
Solvent S1 = Isopropanol-conc. ammonium hydroxide-water (6:3:1)
S2 = *n*-Propanol-conc. ammonium hydroxide-water (6:3:1)
S3 = Chloroform-methanol-conc. ammonium hydroxide-water (12:10:2:1)
Detection D1 = Water
D2 = Rhodamine G viewed with UV light

REFERENCE

Joo, C. N., Park, C. E., Kramer, J. K. G., and Kates, M., *Can. J. Biochem.*, 51, 1527, 1973.

Table TLC 17
TRITERPENES

Layer	L1	L1	L	L1	L1	L1
Solvent	S1	S2	S3	S4	S5	S6
Detection	D1,D2	D1,D2	D1,D2	D1,D2	D1,D2	D1,D2
Compound				**R_F × 100**		
Geraniol	14	24	48	63	72	60
Nerol	15	—	56	65	71	—
Linalool	—	—	81	46	71	—
Myrcene	—	—	100	64	42	—
α-Terpineol	—	—	64	65	76	—
Terpinolene	—	—	100	65	78	—
Eugenol	15	—	78	72	77	73
trans, trans-Farnesol	—	26	51	46	60	52
cis, trans-Farnesol	—	33	60	46	60	52
Pristane	100	100	100	—	—	—
Phytol	—	—	72	—	46	27
Cholesterol	11	22	59	0	45	22
5α-Cholestanol	11	—	57	0	40	17
Lanosterol	20	35	79	0	45	22
24,25-Dihydrolanosterol	20	—	79	0	40	17
Stigmasterol	10	—	51	0	39	—
β-Sitosterol	10	—	51	0	39	20
Cycloartenol	21	—	82	0	39	19
24-Methylenecycloartanol	21	—	82	0	39	20
Cholesteryl acetate	74	—	98	0	—	—
Lanosteryl acetate	77	—	99	0	—	—
Squalene	98	99	100	0	21	7
Squalane	100	100	100	0	—	—
2,3-Oxidosqualene	71	92	—	0	32	—

Layer L1 = 50% Silica-gel G, 43% silica gel H silanized with dimethyldiehlorosilane and 7% $CaSO_4 \cdot {}^1/_2 H_2O$ (w/w/w)

Solvent S1 = Hexane-ethyl acetate (95:5)
S2 = Hexane-ethyl acetate (90:10)
S3 = Hexane-ethyl acetate (80:20)
S4 = Methanol-water (70:30)
S5 = *p*-Dioxane-water (75:25)
S6 = *p*-Dioxane-acetic acid-water (50:30:20)

Detection D1 = Iodine vapor
D2 = 50% aqueous sulfuric acid, heat 120°C

Technique Ascending 14 cm, mixed absorbent permits two-dimensional TLC by absorption and reversed-phase on the same plate

REFERENCE

Vidrine, D. W. and Nicholas, H. J., *J. Chromatogr.*, 89, 92, 1974.

Table TLC 18
ACETYLATED STEROLS

Layer	L1	L2
Solvent	S1	S2
Reference	1	2
Compound	**$R_F \times 100$**	
Acetate of:		
Cycloartanol	—	65
24-Methylenecycloartanol	—	35
Cycloartenol	59	45
24-Methylenecyloartenol	41	—
7-Ene-cholestenol	90	—
Cholestanol	90	—
Sitosterol	84	—
Stigmasterol	82	—
Fucosterol	70	—
Isofucosterol	60	—
24-Methylenecholesterol	55	—
Cholesterol	84	—
Campesterol	84	—
Lanosterol diepoxide	—	20
Cycloartenol monoepoxide	—	40
Cycloartanol epoxide	—	65

Layers L1 = Silica gel H containing 20% (w/w) silver nitrate
L2 = Alumina impregnated with silver nitrate
Solvent S1 = Benzene-hexane (1:1)
S2 = Hexane-ethyl acetate (20:1)
Detection 10% Antimony trichloride in benzene

REFERENCES

1. **Evans, F. J.,** *J. Pharm. Pharmacol.*, 24, 227, 1972.
2. **Alcaide, A., Devys, M., Barbier, M., Kaufmann, H. P., and Sen Gupta, A. K.,** *Phytochemistry*, 10, 209, 1971.

Table TLC 19
PRESQUALENE ALCOHOLS

Layer	L1	L1	L1	L1	L1	L2	L2	L2
Solvent	S1	S2	S3	S4	S5	S6	S7	S8
Compound				**$R_F \times 100$**				
α-*trans*-Presqualene alcohol	28	29	32	26	60	26	61	
β-*trans*-Presqualene alcohol	37	39	48	40	84	34	62	
α-*cis,trans*-Presqualene alcohol	36	35	37	—	—	—	—	
β-*cis,trans*-Presqualene alcohol	38	39	46	—	—	—	—	

Layer L1 = Silica gel GF_{254}
L2 = Silica gel GF_{254} + 10% Ag^+
Solvent S1 = Ethyl ether-carbon tetrachloride (1:4)
S2 = Benzene (3h)
S3 = Ethyl acetate-petroleum ether (1:19; 4.16h)
S4 = Ethyl acetate-petroleum ether (2:23; 3h)
S5 = Isopropylalcohol-benzene (1:95; 2h)
S6 = Ethyl acetate
S7 = Acetone
S8 = Isopropyl alcohol-benzene (1:9), double development.
Detection UV light

REFERENCE

Altman, L. J., Kowerski, R. L., and Laungani, D. R., *J. Am. Chem. Soc.*, 100, 6174, 1978.

Table TLC 20
TRITERPENES

Layer	L1	L1	L1	L1
Solvent	S1	S2	S3	S4
Compound		**$R_F \times 100$**		
Genin G	64	63	70	73
Genin J	55	49	62	67
Genin K	47	38	54	63
Genin N	39	35	50	61
Gymnestrogenin	32	16	44	56
Gymnemagenin	23	10	27	40

Layer L1 = Silica gel G

Solvent S1 = Benzene-methanol-acetic acid (45:8:4)

S2 = Chloroform-methanol (9:1)

S3 = Benzene-chloroform-methanol (5:8:3)

S4 = Benzene-methanol (7:3)

Detection Ceric sulfate-sulfuric acid

REFERENCE

Sinsheimer, J. E. and Subba Rao, G., *J. Pharm. Sci.*, 59, 629, 1970.

Table TLC 21
PENTACYCLIC TRITERPENES

Layer	L1	L1
Solvent	S1	S2
Detection	D1,D2	D1,D2
Compound	**$R_F \times 100$**	
8-β-Glycyrrhetic acid	71	98
3-oxo-Glycyrrhetic acid	84	—
3-*O*-Acetyl-Glycyrrhetic acid	83	—
Carbenoxolone (11 β-Glycyrrhetic acid hemisuccinate)	81	98
Carbenololone 30-*O*-methyl ester	77	—
Carbenoxolone 3-*O*-methyl ether	82	—
Methyl (18-β-Glycyrrhet-30-yl 2,3,4-Triacetyl-β-D-glucopyranosiduronate	—	98
18-β-Glycyrrhet-30-ylβ-D-Glucosiduronic acid	—	60
Carbenoxolone *bis* methyl triacetyl glucosiduronate	—	98
3-*O*-β(3-Carboxypropionyl)-18-β-glycyrrhet-30-yl-β-D-glucosiduronic acid	—	62

Layer L1 = Silica gel HF_{254}
Solvents S1 = Ethyl acetate
S2 = Acetic acid-1,2-dichloroethane-butanol-water (4:4:1:1)
Detection D1 = 5% Antimony pentachloride in chloroform; heat at 120°C for 10 min to give brown spots
D2 = 0.2% Napthoresorcinol in acetone and 9% aqueous phosphoric acid; heat 120°C for 10 min glucosiduronic acid to give black spots

REFERENCE

Iveson, P. and Parke, D. V., *J. Chem. Soc. (C),* 2038, 1970.

Table TLC 22
CAROTENOIDS

Layer	L1	L1	L1	L1	L2	L2	L2
Solvent	S1	S2	S3	S4	S5	S6	S7
Detection	D1-D3	D1-D3	D1-D3	D1-D3	D1-D3	D1-D3	D1-D3
Compound	$R_F \times 100$						
Squalene	16	53	45	65	58	68	73
4,4′-Diapophytoene	9	46	34	55	45	61	67
Phytoene	4	35	23	47	35	56	62
4,4′-Diapophytofluene	7	35	25	45	19	38	58
Phytofluene	1	28	17	34	11	32	51
4,4′-Diapo-7,8,11,12-tetrahydrolycopene	2	26	19	36	7	18	37
4,4′-Diapo-ζ-carotene	2	25	19	35	7	16	37
ζ-Carotene	0	23	11	28	3	10	35
4,4′-Diaponeurosporene	0	22	14	27	0	4	21
Neurosporene	0	18	9	21	0	0	18

Layers L1 = Silica gel G
L2 = Alumina G
Solvent S1 = Light petroleum (40-60°C)
S2 = 1% Diethyl ether in light petroleum
S3 = 5% Benzene in light petroleum
S4 = 10% Benzene in light petroleum
S5 = 0.25% Diethyl ether in hexane
S6 = 1% Diethyl ether in hexane
S7 = 5% Benzene in hexane
Detection D1 = Visible light
D2 = Fluorescence by UV light (360 nm)
D3 = Iodine
Technique Tank presaturated 1 hr before use

REFERENCE

Taylor, R. F. and Davies, B. H., *Biochem. J.*, 139, 751, 1974.

Table TLC 23
CAROTENOIDS

Layer	L1	L1	L1	L2	L2
Solvent	S1	S2	S3	S1	S4
Detection	D1-D3	D1-D3	D1-D3	D1-D3	D1-D3
Compound			$R_F \times 100$		
Squalene	41	67	78	24	53
Dehydrosqualene	26	—	—	10	40
Phytoene	19	46	67	10	40
cis-Phytofluene	—	17	40	10	40
trans-Phytofluene	—	11	32	10	40
neo-x-Carotene	—	5	16	10	40
β-Carotene	—	3	8	10	40
neo-β-Carotene	—	3	8	10	40
Standards					
Squalene	41	67	78	24	54
Phytoene	19	46	67	10	40
cis-Phytofluene	—	17	42	10	40
trans-Phytofluene	—	11	31	10	40
x-Carotene	—	5	16	10	40
β-Carotene	—	3	7	10	40

Layers L1 = Alumina G (Type E)
L2 = Silica gel H

Solvent S1 = Ethyl ether-hexane (0.25:99.75)
S2 = Ethyl ether-hexane (1:99)
S3 = Ethyl ether-hexane (3:97)
S4 = Ethyl ether-hexane (0.5:99.5)

Detection D1 = Iodine vapor
D2 = UV and visible light
D3 = conc. H_2SO_4, heat

REFERENCE

Kushwaha, S. C., Rugh, E. L., Kramer, J. K. G., and Kates, M., *Biochim. Biophys. Acta,* 260, 492, 1972.

Table TLC 24
CAROTENOIDS

Layer	L1	L1	L1	L2	L2
Solvent	S1	S2	S3	S4	S2
Compound			**$R_F \times 100$**		
β,ε-Carotene	71	—	—	69	—
β,β-Carotene	71	—	—	57	—
β,ψ-Carotene	64	—	—	46	—
Lycopene	57	—	—	3	—
2,2′-Diol	—	67	—	—	52
Lutein	—	61	—	—	48
Zeaxanthin	—	61	—	—	29
all-*trans* Lutein	—	—	70	—	—
Neo U Lutein	—	—	62	—	—
Neo V Lutein	—	—	52	—	—
Neo A Bacterioruberin	—	—	61	—	—
All-*trans*-Bacterioruberin	—	—	41	—	—
Lutein 3′-ether	—	54	—	—	—
Epimer 1 Lutein 3′-ether	—	54	—	—	—
Epimer 2 Lutein 3′-ether	—	54	—	—	—
Epimer 1′ Auroxanthin	—	20	—	—	—
Epimer 2 Auroxanthin	—	20	—	—	—
Epimer 1 Neochrome	—	—	60-63	—	—
Epimer 2 Neochrome	—	—	60-63	—	—
Mono-*cis* Neochrome	—	—	60-63	—	—

Layer L1 = Silica gel
L2 = Silica gel G - $Ca(OH)_2$-MgO - $CaSO_4$ (10:4:3:1 w/w)

Solvent S1 = Acetone-hexane (2:98)
S2 = Acetone-hexane (40:60)
S3 = Acetone-hexane (60:40)
S4 = Acetone-hexane (5:95)

Detection UV and visible light

REFERENCE

Fiksdahl, A., Mortensen, J. T., and Liaaen-Jensen, S., *J. Chromatogr.*, 157, 111, 1978.

Table TLC 25
CAROTENOIDS

Layer		L1	L1	L1	L1	L1	L1	L1	L1	L1
Solvent		S1	S2	S3	S4	S5	S6	S7	S8	S9
Compound	**Structural features[a]**					**$R_F \times 100$**				
Physalien	2COOR	100	100	100	95	30	0	0	0	0
β-Carotene	—	100	100	100	100	80	25	10	5	0
Lycopene	2 ψ	95	95	90	75	60	5	0	0	0
β-Cryptoxanthin	OH	57	62	70	76	74	39	21	9	4
Rubixanthin	ψ, OH	36	45	60	64	45	15	4	0	0
Lycoxanthin	2ψ, OH	16	29	37	40	32	8	0	0	0
Lutein	ε, 2OH	16	35	57	93	95	68	45	24	14
Zeaxanthin	2OH	12	30	54	78	82	55	35	18	10
Isozeaxanthin	2OH	16	34	56	92	91	57	36	19	10
Eschscholtzxanthin	*R*, 2OH	0	12	22	25	22	8	1	0	0
Lycophyll	2ψ, 2OH	0	8	20	22	20	7	0	0	0
Violaxanthin	2OH, 2E	11	30	52	76	93	88	73	55	35
Neoxanthin	3OH, E	2	22	42	82	96	90	78	60	40
Crustaxanthin	4OH	5	18	33	86	95	81	62	44	33
Echinonome	C=O	90	91	92	90	72	35	18	6	2
3-Oxoechinenone	2C=O	59	62	68	81	80	54	34	17	9
Canthaxanthin	2C=O	55	58	65	79	80	55	37	20	10
Rhodoxanthin	*R*, 2CO=O	14	28	42	43	40	14	7	2	1
Astacene	4C=O	25	34	50	69	72	42	25	12	7
Capsanthin	C=O, 20H	4	24	42	79	81	62	42	25	15
Capsorubin	2C=O, 2OH	10	19	37	74	76	60	42	25	15
Fucoxanthin	C=O, 2OH,OR,E	7	27	44	84	98	95	85	75	55
β-Apo-8′-carotenoic acid	COOH	13	28	38	30	15	5	0	0	0
Torularhodin	ψ, COOH	0	6	10	9	2	1	0	0	0

[a] Carotenoids have β-end groups except where stated otherwise; *E=E* end group ψ = ψ end group, R = retro, E = 5,6-epoxide.

Layer	L1 = Polyamide (Merck)-cellulose MN 300 (85:15) slurried in methanol air dried (30 min)
Solvent	S1 = Light petroleum-methanol-methyl ethyl ketone (20:1:1)
	S2 = Light petroleum-methanol-methyl ethyl ketone (8:1:1)
	S3 = Light petroleum-methanol-methyl ethyl ketone (4:1:1)
	S4 = Light petroleum-methanol-methyl ethyl ketone (2:1:1)
	S5 = Methanol-methyl ethyl ketone (1:1)
	S6 = Methanol-methyl ethyl ketone-water (5:5:1)
	S7 = Methanol-methyl ethyl ketone-water (3:3:1)
	S8 = Methanol-methyl ethyl ketone-water (2:2:1)
	S9 = Methanol-methyl ethyl ketone-water (3:3:2)
Detection	Visible light

REFERENCE

Egger, K. and Voigt, H., *Z. Pflanzenphysiol.*, 54, 407, 196.

Compiled from **Davies, B. H.,** in *Chemistry and Biochemistry of Plant Pigments*, Vol. 2, 2nd ed., Goodwin, T. W., Ed., Academic Press, London, 1976, 102.

Table TLC 26
CAROTENOIDS

Layer	L1	L1	L1	L1	L1	L2	L3
Solvent	S1	S2	S3	S4	S5	S6	S7
Compound				**R_F × 100**			
β-Carotene	95	98	98	98	—	10	0
γ-Carotene	68	—	—	—	—	15	0
Neurosporene	66	—	—	—	—	—	—
Lycopene	53	68	—	—	—	—	—
β-Cryptoxanthin	29	62	81	91	—	90	7
Torulene	25	—	—	—	—	—	—
Plectaniaxanthin diester	—	47	80	—	—	—	—
3,4-Dehydrotorulene	—	35	—	—	—	—	—
Torularhodin methyl ester	—	—	58	90	—	48	—
2′-Dehydroplectaniaxanthin ester	—	—	57	90	—	—	—
3,4-Dehydrolycopene	—	—	52	—	—	—	—
2′-Dehydroplectaniaxanthin	—	—	40	75	—	—	—
Lutein	—	—	39	72	91	—	56
Zeaxanthin	—	9	30	59	87	—	—
Torularhodinaldehyde	—	—	29	—	—	—	—
Astaxanthin	—	—	—	57	85	—	—
Plectaniaxanthin	—	—	—	45	—	—	—
3′,4′-Didehydro-β,ψ-carotene-16′-ol	—	—	19	—	—	—	—
Violaxanthin	—	—	18	44	83	—	84
Fucoxanthin	—	—	—	40	81	—	—
Phillipsiaxanthin diester	—	—	—	35	—	—	—
Rhodoxanthin	—	—	10	27	72	—	26
Phillipsiaxanthin	—	—	—	—	36	—	—
Torularhodin	—	—	—	10	—	—	—
Lutein dipalmitate (helenien)	—	—	—	—	—	2	0
Isozeaxanthin dimethyl ether	—	—	—	—	—	60	—
Echinenone	—	—	—	—	—	61	—
Isozeaxanthin	—	—	—	—	—	—	49
Zeaxanthin	—	—	—	—	—	—	54
Taraxanthin	—	—	—	—	—	—	72
Capsanthin	—	—	—	—	—	—	74
Neoxanthin	—	—	—	—	—	—	95

Layer
L1 = Kieselguhr-loaded paper Schleicher and Schüll no. 287
L2 = Kieselguhr impregnated with liquid paraffin (7% in light petroleum)
L3 = Kieselguhr impregnated with vegetable oil (7% palmin-livio (1:1) in light petroleum)

Solvent
S1 = Light petroleum
S2 = Light petroleum-acetone (98:2)
S3 = Light petroleum-acetone (95:5)
S4 = Light petroleum-acetone (90:10)
S5 = Light petroleum-acetone (80:20)
S6 = Methanol-acetone (5:2)
S7 = Methanol-acetone-water (20:4:3)

Detection UV and visible light

REFERENCE

Compiled by **Davies, B. H.**, in *Chemistry and Biochemistry of Plant Pigments*, Vol. 2, 2nd ed., Goodwin, T. W., Ed., Academic Press, London, 1976, 83.

Table TLC 27
XANTHOPHYLLS

Layer	L1	L1	L2	L2	L3	L4	L5
Solvent	S1	S2	S3	S4	S2	S5	S6
Compound				$R_F \times 100$			
Torularhodin methyl ester	83	—	100	—	—	—	—
β-Apo-12′-carotenal	70	—	—	—	—	—	—
β-Apo-8′-carotenal	64	—	—	—	—	—	—
Apo-8′-lycopenal	58	—	—	—	—	—	—
β-Apo-10′-carotenal	53	—	—	—	—	—	—
Methyl bixin	13	81	97	—	—	—	—
Echinenone	—	—	—	90	82	88	90
β-Carotene 5,6,5′, 6′-diepoxide	—	—	—	—	—	86	—
Canthaxanthin	—	63	90	82	43	83	35
Hydroxy-α-carotene	—	—	—	72	—	—	—
β-Cryptoxanthin	—	54	75	70	34	77	58
Cryptocapsin	—	—	—	64	—	—	—
Rhodoxanthin	—	—	—	—	16	—	33
Lutein	—	10	35	57	—	65	—
Zeaxanthin	—	5	24	53	—	39	10
Taraxanthin	—	—	—	—	—	75	—
Crocoxanthin	—	—	—	—	—	75	—
Fucoxanthin	—	—	—	—	—	72	—
Antheraxanthin	—	10	32	40	—	52	—
Violaxanthin	—	—	21	35	—	74	—
Capsanthin	—	5	16	29	—	—	—
Capsanthin 5,6-epoxide	—	—	—	24	—	—	—
Capsorubin	—	5	13	20	—	—	—
Neoxanthin	—	—	—	15	—	49	—
Diadinoxanthin	—	—	—	—	—	41	—
Vaucheriaxanthin	—	—	—	—	—	19	—
Heteroxanthin	—	—	—	—	—	12	—
Oscillaxanthin	—	—	—	—	—	6	—
Aphanizophyll	—	—	—	—	—	5	—
Myxoxanthophyll	—	—	—	—	—	3	—
Bixin	—	—	5	—	—	—	—
Azafrin	—	—	2	—	—	—	—

Layer
L1 = Sec. magnesium phosphate (activated)
L2 = Silica gel G (activated)
L3 = Calcium hydroxide-silica gel G (6:1 w/w, activated)
L4 = Calcium carbonate-magnesium oxide-calcium hydroxide (30:6:4 by wt, activated)
L5 = MN 300 cellulose

Solvent
S1 = Carbon tetrachloride
S2 = Benzene
S3 = Dichloromethane-ethyl acetate (4:1)
S4 = Benzene-ethyl acetate-methanol (75:20:5)
S5 = Light petroleum-acetone-chloroform-methanol (50:50:40:1)
S6 = Light petroleum-carbon tetrachloride (5:2)

Detection Visible light

Table TLC 27 (continued)
XANTHOPHYLLS

REFERENCE

Compiled by **Davies, B. H.**, in *Chemistry and Biochemistry of Plant Pigments*, Vol. 2, 2nd ed., Goodwin, T. W., Ed., Academic Press, London, 1976, 78.

Table TLC 28
CAROTENOIDS

Layer	L1	L1	L1	L1
Solvent	S1	S2	S3	S4
Compound		**$R_F \times 100$**		
α-Carotene	43	74	70	82
β-Carotene	29	74	70	82
Lycopene	2	75	5	8
Crocin	0	0	0	0
Crocetin	0	0	0	0
Bixin	0	0	0	0

Layer L1 = Magnesium hydroxide
Solvent S1 = Carbon disulfide
S2 = Pyridine
S3 = Tetrahydrofuran
S4 = Ethanol
Detection UV light

REFERENCE

Keefer, L. K. and Johnson, D. E., *J. Chromatogr.*, 69, 215, 1972.

Table TLC 29
APOCAROTENALS

Layer	L1	L1	L1	L1
Solvent	S1	S2	S3	S4
Compound		**$R_F \times 100$**		
β-Apo-8′-carotenal	67	67	48	—
β-Apo-10′-carotenal	58	53	36	—
β-Apo-12′-carotenal	71	63	50	—
5,8-Epoxy-β-apo-8′-carotenal	50	39	29	—
5,8-Epoxy-β-apo-12′-carotenal	55	42	32	—
β-Apo-8′-carotenyl acetate	77	78	60	—
5,8-Epoxy-β-apo-8′-carotenyl acetate	61	56	38	—
β-Apo-8′-carotenoic acid	0	0	0	22
β-Apo-10′-carotenoic acid	0	0	0	12
β-Apo-12′-carotenoic acid	3	3	3	28
5,8-Epoxy-β-apo-8′-carotenoic acid	4	5	0	33
5,8-Epoxy-β-apo-12′-carotenoic acid	5	7	3	42
β-Carotene	100	100	100	—

Layer L1 = Silica gel-calcium sulfate (8:1)
Solvent S1 = Acetone-light petroleum (10:90)
S2 = Diethyl ether-light petroleum (25:75)
S3 = Acetone-cyclohexane (6:94)
S4 = Acetone-light petroleum (25:75)
Detection UV light and $SbCl_3$
Technique Ascending development

REFERENCE

Singh, H., John, J., and Cama, H. R., *J. Chromatogr.*, 75, 146, 1973.

Table TLC 30
RETINOIDS

Layer	L1	L1	L1	L2	L3
Solvent	S1	S2	S3	S4	S5
Reference	1	1	1	2	3
Compound			**$R_F \times 100$**		
Retinol	29	5	3	—	26
Retinal	61	22	15	65	40
Retinyl acetate	75	48	39	—	54
Retinoic acid	—	—	—	8	21
3-Dehydroretinoic acid	—	—	—	8	—
3-Dehydroretinol	—	—	—	61	—
Retinyl palmitate	—	—	—	—	71

Layer L1 = Magnesium hydroxide
L2 = Silica gel-$CaSO_4$ (9:1)
L3 = Silica gel 60 F_{254}

Solvent S1 = Benzene
S2 = Carbon tetrachloride
S3 = Carbon disulfide
S4 = Acetone-light petroleum (10:90)
S5 = Acetone-light petroleum (18:82)

Detection UV light or antimony pentachloride

REFERENCES

1. **Keefer, L. K. and Johnson, D. E.,** *J. Chromatogr.*, 69, 215, 1972.
2. **Singh, H., John, J., and Cama, H. R.,** *J. Chromatogr.*, 75, 146, 1973.
3. **Fung, Y. K. and Rahwan, R. G.,** *J. Chromatogr.*, 147, 528, 1978.

Table TLC 31
ISOPRENOID QUINONES

Layer	L1	L1	L2	L3
Solvent	S1	S2	S3	S4
Compound		**$R_F \times 100$**		
6-Methoxy-2-heptaprenyl-1,4-benzoquinone	40	10	—	—
6-Methoxy-2-octaprenyl (H_2)-1,4-benzoquinone	40	10	—	50
6-Methoxy-2-nonaprenyl-1,4-benzoquinone	—	—	—	43
6-Methoxy-2-decaprenyl-1,4-benzoquinone (X-H_2)	—	—	—	29
5-Demethoxyubiquinone-6	—	—	—	60
5-Demethoxyubiquinone-7	—	—	—	54
5-Demethoxyubiquinone-8	—	—	—	47
5-Demethoxyubiquinone-9	—	—	—	38
5-Demethoxyubiquinone-9(H_2)	45	15	—	29
5-Demethoxyubiquinone-10	—	—	—	29
5-Demethoxyubiquinone-10(H_2)	45	15	—	22
Ubiquinone-6	—	—	—	53
Ubiquinone-7	—	—	—	45
Ubiquinone-7(H_2)	—	—	—	37
Ubiquinone-8	—	—	52	37
Ubiquinone-8(H_2)	—	—	55	29
Ubiquinone-9	—	—	50	29
Ubiquinone-9(H_2)	—	—	53	20
Ubiquinone-10	—	—	48	20
Ubiquinone-10(X-H_2)	53	23	51	11
Ubiquinone-11	53	23	46	11

Layer L1 = Silica-gel G
L2 = Silica-gel G impregnated with silver nitrate
L3 = Silica-gel G impregnated with paraffin
Solvent S1 = Benzene-chloroform (1:1)
S2 = Benzene
S3 = Acetone-butanone (1:4)
S4 = Water-acetone (1:9)
Detection Rhodamine 6G and UV light

REFERENCE

Law, A., Threlfall, D. R., and Whistance, G. R., *Biochem. J.*, 123, 331, 1971.

Table TLC 32
UBIQUINONES

Layer	L1	L1	L2	L2
Solvent	S1	S2	S3	S4
Detection	D1-D4	D1-D4	D1-D4	D1-D4
Compound		$R_F \times 100$		
Ubiquinol-4	78	—	—	—
Ubiquinol-5	74	—	—	—
Ubiquinol-6	70	—	—	—
Ubiquinol-7	65	—	—	—
Ubiquinol-8	60	—	—	—
Ubiquinol-9	55	—	—	—
Ubiquinol-10	49	—	—	—
Ubiquinone-8	—	—	—	52
Ubiquinone-9	—	—	—	50
Ubiquinone-10	—	—	—	48
6-Methoxy-2-octaprenylphenol	—	50	—	—
Demethylmenaquinone-8	—	40	56	—
Menaquinone-8	—	30	56	—

Layer L1 = Silica gel G impregnated with paraffin
L2 = Silica gel G impregnated with silver nitrate
Solvent S1 = Water-acetone (3:17)
S2 = Water-acetone (1:19)
S3 = Acetone-butan-2-one (1:19)
S4 = Acetone-butan-2-one (1:4)
Detection D1 = 0.1% Ferric chloride + 0.05% α,α′-bipyridyl in water-ethanol (1:1)
D2 = Rhodamine 6G and UV light
D3 = 0.002% Sodium fluorescein in ethanol and UV light
D4 = Gibbs' reagent

REFERENCE

Whistance, G. R., Dillon, J. F., and Threlfall, D. R., *Biochem. J.*, 111, 461, 196

Table TLC 33
MENAQUINONES

Layer	L1	L1	L2	L3	L3	L4	L4	L1	L3	L4
Solvent	S1	S2	S3	S4	S2	S4	S2	S5	S6	S7
Detection	D1-D3	D1-D3	D1-D3	D1-D3	D1-D3	D1-D3	D1-D3	na	na	na
Reference	1	1	1	1	1	1	1	2	2	2
Compound					**$R_F \times 100$**					
Phylloquinone	43	70	71	—	—	94	72	35	38	97
Menaquinone-4	35	60	85	37	25	33	25	29	66	48
Menaquinone-6	39	66	73	35	22	25	19	—	—	—
Menaquinone-9	46	80	40	33	21	13	12	—	—	—
Menaquinone-9(H)	46	81	36	40	29	25	18	—	—	—
Menaquinone-10	45	82	32	26	19	12	9	—	—	—
2,3-Epoxymenaquinone-4	—	—	—	—	—	—	—	27	81	46
Demethylmenaquinone-4	—	—	—	—	—	—	—	27	72	45
Menaquinone-3	—	—	—	—	—	—	—	28	79	67
2,3-Epoxymenaquinone-3	—	—	—	—	—	—	—	27	89	59
2,3-Epoxymenaquinone	—	—	—	—	—	—	—	33	59	91
Menadione	—	—	—	—	—	—	—	12	96	62

Layers
L1 = Silica gel G
L2 = Silica gel G dipped in 5% paraffin in hexane
L3 = Paraffin impregnated kieselguhr (dipped in 5% liquid paraffin in light petroleum)
L4 = 10% (w/w) Silver nitrate-silica gel G

Solvents
S1 = Heptane-benzene (1:1)
S2 = Heptane-benzene (2:8)
S3 = Acetone-water (96:4)
S4 = Benzene
S5 = Diethyl ether-light petroleum (6:94)
S6 = Water-acetone (10:90)
S7 = Diisopropyl ether-light petroleum (60:40)

Table TLC 33 (continued)
MENAQUINONES

Detection D1 = UV light
D2 = 2′,7′-Dichlorofluorescein
D3 = Rhodamine B

REFERENCES

1. **Matschiner, J. T. and Amelotti, J. M.,** *J. Lipid Res.*, 9, 176, 1968.
2. **Donnahey, P. L., Burt, V. T., Rees, H. H., and Pennock, J. F.,** *J. Chromatogr.*, 170, 272, 1979.

Table TLC 34
NAPHTHOQUINONES AND VITAMINS

Layer	L1	L1	L2
Solvent	S1	S2	S3
Compound		**R_F × 100**	
1,2-Naphthoquinone	10	48	25
1,4-Naphthoquinone	28	61	54
Phylloquinone	61	80	75
Menadione	30	65	59
Mendiol diacetate	13	59	41
Calciferol	8	53	30
Retinyl acetate	20	52	47
Cholesterol	9	46	28

Layer L1 = Silica gel
L2 = Silica gel with 0.25 mℓ polyethylene glycol 200 per 200 g

Solvent S1 = Benzene
S2 = Benzene-methyl ethyl ketone (3:1)
S3 = Benzene-light petroleum (1:1)

Detection 70% Perchloric acid, heat 5-10 min at 105°C, olive green or violet colors

REFERENCE

Rittich, B., Simek, M., and Coupek, J., *J. Chromatogr.*, 133, 345, 1977.

Table TLC 35
MENAQUINONES

Layer	L1	L1	L2
Solvent	S1	S2	S3
Compound		**R_F × 100**	
MK-7	44	—	—
MK-8	40	26[a]	42
MK-9	34	21[a]	31
MK-10	28	15[a]	26
MK-9 (2H)	—	—	62
MK-9 (4H)	24	20[a]	67

[a] Perhydroderivatives of menaquinones were used.

Layer — L1 = Paraffin impregnated silica gel F 254 (5% paraffin oil in petroleum ether)
L2 = Silver nitrate treated silica gel H (25% silver nitrate in acetonitrile)

Solvent — S1 = Acetone-water (95:5)
S2 = Acetone-methylethylketone (3:2)
S3 = Heptane-acetone (4:1)

Detection — H_2SO_4, heat at 150°C

REFERENCE

Sone, N., *J. Biochem. (Tokyo),* 76, 133, 1974.

Table TLC 36
TOCOPHEROLS

Layer	L1
Solvent	S1

Compound	$R_F \times 100$
d-δ-Tocopherol	16.5 ± 3.4
d-γ-Tocopherol	25.5 ± 2.8
d-β-Tocopherol	28.7 ± 3.4
d-α-Tocopherol	36.2 ± 5.4
Retinol acetate	63.5 ± 7.4

Layer L1 = Silica gel G
Solvent S1 = Cyclohexane-n-hexane-isopropylether-ammonium hydroxide (40:40:20:2)
Detection Combination of phosphomolybdic acid-dichlorofluorescein, inspect with UV light (254 nm)
Technique Ascending

REFERENCE

Lovelady, H. G., *J. Chromatogr.*, 78, 449, 1973.

Section II
Detection Reagents for Thin Layer Chromatography of Terpenes

DETECTION REAGENTS FOR TLC

INTRODUCTION

Listed in this section are various reagents that have been used to detect terpenoids on thin layer plates. The section begins with a categorization of the reagents according to the specific terpenoids that they are capable of visualizing. The reagents are arranged under the various terpene headings in the same sequence as the chromatography tables. There are preparative procedures for the various reagents given. Methods and typical results are also described. In the latter subsection the reagents are listed alphabetically.

MONOTERPENES

2. Anisaldehyde-acetic acid-methanol-sulfuric acid
5. Anisaldehyde-methanol-sulfuric acid
6. Antimony pentachloride
13. Chlorosulfonic acid-acetic acid (unsaturated cyclic hydrocarbons)
24. Fast blue (tetrahydrocannabinoids and other phenols)
35. Phosphomolybdic acid
37. Potassium permanganate
43. Sodium methoxide-dimethyl sulfoxide
47. Vanillin-sulfuric acid

SESQUITERPENES

3. Anisaldehyde-acetic acid-sulfuric acid
6. Antimony pentachloride
10. Ceric sulfate-sulfuric acid
14. Cinnamaldehyde
12. Chloramine-trichloroacetic acid
13. Chlorosulfonic acid-acetic acid (unsaturated cyclic hydrocarbons)
11. Copper acetate-phosphoric acid
18. Dimethylaminobenzaldehyde
19. Dinitrobenzoic acid
33. Orthophosphoric acid
35. Phosphomolybdic acid
45. Toluenesulfonic acid
46. Trichloroacetic acid

DITERPENES

3. Anisaldehyde-acetic acid-sulfuric acid
13. Chlorosulfonic acid-acetic acid (unsaturated cyclic hydrocarbons)
27. Godin's reagent
31. Methanol-sulfuric acid
47. Vanillin-sulfuric acid

TRITERPENES

1. Anisaldehyde-antimony trichloride
6. Antimony pentachloride
10. Ceric sulfate-sulfuric acid
12. Chloramine-trichloracetic acid
30. Liebermann-Burchard reagent
45. Toluenesulfonic acid
46. Trichloroacetic acid

CAROTENOIDS

7. Antimony trichloride
21. 2-Diphenyl-indane-1,3-dione-1-hydrazine
28. Hydrogen chloride
29. Iodine vapor
40. Rhodamine-6G
41. Rhodamine-ammonium hydroxide

POLYPRENOLS

4. Anisaldehyde-ethanol-sulfuric acid
1. Anisaldehyde-antimony trichloride
6. Antimony pentachloride
17. Dichlorofluorescein
25. Fluorescein
33. Orthophosphoric acid
35. Phosphomolybdic acid
40. Rhodamine-6G

PHOSPHORYLATED ISOPRENOIDS

42. Rosenberg's reagent

RETINOIDS

7. Antimony trichloride

TOCOPHEROLS

6. Antimony pentachloride
8. Bathophenanthroline
10. Ceric sulfate-sulfuric acid
15. *o*-Dianisidine
17. Dichlorofluorescein
23. Emmerie-Engel reagent
35. Phosphomolybdic acid
36. Phosphomolybdic acid-2′,7′-dichlorofluorescein

VITAMIN K

34. Perchloric acid
25. Fluorescein
32. Neotetrazolium

UBIQUINONES

9. α,α′-Bipyridyl-ferric chloride (for ubiquinols)

GENERAL USE

4. Anisaldehyde-ethanol-sulfuric acid (for terpene alcohols)
10. Ceric sulfate-sulfuric acid
11. Copper acetate-phosphoric acid
17. Dichlorofluorescein
20. Dinitrophenylhydrazine (for oxo compounds in general)
22. 2,2-Diphenyl-1-picrylhydrazyl
25. Fluorescein
26. Gibbs' reagent (polyprenyl phenols)
29. Iodine vapor
37. Potassium permanganate
39. Rhodamine-B
40. Rhodamine-6G
44. Sulfuric acid

1. Anisaldehyde-Antimony Trichloride

Preparation: Add anisaldehyde (1 mℓ) to 100 mℓ chloroform saturated with antimony trichloride followed by 2 mℓ conc. sulfuric acid, forming a two-phase system. Keep this solution in the dark at room temperature for 1.5 hr before use. Can be used for several days if refrigerated.

Procedure: Spray the upper phase and heat the plate at 90°C for 3 min. Spots are detected by visible and UV light.

Remarks: Characteristic colors are obtained for triterpenes and polyprenols.

2. Anisaldehyde-Acetic Acid-Methanol-Sulfuric Acid

Preparation: Dissolve 0.5 mℓ anisaldehyde in 10 mℓ acetic acid and 85 mℓ methanol Add 5 mℓ conc. sulfuric acid.

Procedure: Spray plates and heat to 100 to 110°C for 3 to 7 min.

Results: Monoterpene ketones give characteristic colors

Menthone	by	Piperitone	y-o
Pulegone	gy-bn	Piperitenone	y-o
Dihydrocarvone	r-bn	Carvenone	r-bn
Carvone	bn-vt		

3. Anisaldehyde-Acetic Acid-Sulfuric Acid

Preparation: Add 1 mℓ anisaldehyde to 97 mℓ of glacial acetic acid then add 2 mℓ conc. sulfuric acid.

Procedure: Spray plates and heat at 120°C for 6 min.

Results: Sesquiterpene lactones (at room temperature) and diterpene acetates (phorbols) give characteristic colors. After heating, virtually all, unsaturated or oxygenated sesquiterpenes are detected.

Sesquiterpenes	Daylight	UV (360 nm)
Gafrinin	-	light y
Dihydrogriesenin	-	light bl
Griesen	-	o

Diterpenes	Daylight	UV (360 nm)
Phorbol triacetate	pk	y-bn
12-Deoxyphorbol diacetate	r-bn	dull-r
4-Deoxy-4α-phorbol triacetate	y-bn	light-y
4α-Phorbol triacetate	bn	o
4α-Phorbol tetraacetate	ol-bn	y-bn
Crotophorbolone monoacetate	pk	o
Ingenol triacetate	ol-bn	light-y

4. Anisaldehyde-Ethanol-Sulfuric Acid

Preparation: Add 0.5 g anisaldehyde and 0.5 mℓ conc. sulfuric acid to 9 mℓ of ethanol.

Procedure: Spray chromatograms and heat at 90 to 100°C for 5 to 10 min.

Results: Detects terpene alcohols.

Geraniol	b
trans-trans-Farnesol	p
Geranyl geraniol	mv
Linalool	gn
Nerolidol	bn-y
Geranyl linalool	y-bn

5. Anisaldehyde-Methanol-Sulfuric Acid

Preparation: Anisaldehyde-methanol-conc. sulfuric acid (1:50:1)

Procedure: Spray plates and heat at 100 to 110°C for 5 min.

Results: Detects oxygenated limonene and its derivatives, which give a blue to purple spot on a white background.

6. Antimony Pentachloride

Preparation: 20% Antimony pentachloride in carbon tetrachloride or 5 to 20% antimony pentachloride in chloroform.

Procedure: Spray plates and heat at 100 to 110°C for 3 to 7 min.

Results: Monoterpene ketones give characteristic colors. Also detects sesquiterpenes, triterpenes, polyprenols, tocopherols, and retinoids.

Menthone	gy-bn
Dihydrocarvone	r-bn
Pulegone	bn
Carvone	r-bn
Piperitone	bn
Carvenone	y-bn
Piperitenone	y

7. Antimony Trichloride

Preparation: 20% Antimony trichloride in chloroform (free from ethanol by elution through an activated alumina column).

Procedure: Spray plates and heat at 100°C for 10 min.

Results: Detects carotenoids and A vitamers.

Comment: A vitamers give blue spots, their epoxy derivatives are yellow. Gafrinin (sesquiterpene)-p.

8. Bathophenanthroline

Preparation: Mix equal parts of 0.5% (w/v) bathophenanthroline and 0.2% ferric chloride in absolute ethanol.

Procedure: Spray fluor-containing plates; Scan with UV light at 254 nm.

Results: Detects tocopherols. Higher sensitivity is achieved using plates containing sodium fluorescein instead of commercially available silica gel containing fluorescein.

9. α,α′-Bipyridyl-Ferric Chloride

Preparation: 0.05% αα′-Bipyridyl and 0.1% ferric chloride in ethanol-water (1:1).

Procedure: Spray fluor-containing plates and scan with UV light (254 nm).

Results: Detects ubiquinols, tocopherols.

10. Ceric Sulfate-Sulfuric Acid

Preparation: Triturate 0.1 g ceric sulfate in 4 mℓ water. Add 1 g trichloroacetic acid, boil, add 1.84 mℓ conc. sulfuric acid dropwise until the solution clarifies.

Procedure: Spray plates and heat at 110°C.

Results: Detects sesquiterpene lactones, tocopherols, sterols.

	Daylight	UV (360 nm)		Day light	UV (360 nm)
Gafrinin	p	o	Vermeein	w	light bl
Geigerin	w	bl	Dihydrogriesenin	bn	bl-p
Geigerin	light y	y	Griesenin	bn	y

Comments: A modified preparation (0.1 g ceric sulfate in 10 mℓ 30% sulfuric acid) detects β-tocopherol-br, γ-tocopherol-bl.

11. Copper Acetate-Phosphoric Acid

Preparation: To 2 g $Cu_2(OAc)_4{\cdot}(H_2O)_2$ in 20 mℓ of water, add 20 mℓ of 85% H_3PO_4, and then dilute to 100 mℓ with water.

Procedures: Spray chromatogram and heat to 120°C for 5 min preferably on a metal hot plate observing colors as they develop.

Results: Most unsaturated or oxygenated terpenoids give grey to black spots (5 to 15 μg detectable) eventually. Characteristic colors develop upon initial heating: olefins and alcohols, lavender, blue, or pink; conjugated enones, yellow, green, or red.

Saturated monoterpene ketones (and alcohols) give only feeble spots.

12. Chloramine-Trichloroacetic Acid

Preparation: (A) 3% Chloramine (freshly prepared); (B) 25% Trichloroacetic acid (prepared every few days). Mix 10 mℓ of A with 40 mℓ of B before use.

Procedure: Spray plates and heat for 7 min at 110°C. Irradiate with filtered UV light.

Results: Detects sesquiterpene lactones and sterol glycosides as yellow spots. Sterols give blue spots.

13. Chlorosulfonic Acid-Acetic Acid

Preparation: Chlorosulfonic acid-acetic acid (1:2).

Procedure: Spray chromatogram and heat for 10 min at 130°C.

Results: Detects cyclic unsaturated hydrocarbons.

Comments: Colors: humulene-bn; caryophyllene-bl; longifolene-pk.

14. Cinnamaldehyde

Preparation: 5 mℓ Cinnamaldehyde in ethanol-conc. HCl (95:5). Prepare fresh before use.

Procedure: Spray chromatogram.

Results: Sesquiterpene lactones give colored spots.

15. *o*-Dianisidine

Preparation: Saturated *o*-dianisidine in glacial acetic acid.

Procedure: Spray chromatogram and heat at 120°C for several min.

Results: α,β, and γ-tocopherols give blue violet spots whereas δ gives indigo. ϵ-isomer can also be distinguished. Simpler than many other methods. Oxo compounds also detected.

17. Dichlorofluorescein

Preparation: 0.2% 2,′7′-Dichlorofluorescein in ethanol.

Procedure: Spray plates and irradiate with short wave UV light.

Results: Most saturated and unsaturated lipids give yellow-green fluorescent spots. Detects tocopherol.

Comments: Can also be used in layers either alone or together with Rhodamine B. Does not work on plates treated with silicones.

18. *p*-Dimethylaminobenzaldehyde

Preparation: Dissolve 0.25 g *p*-dimethylaminobenzaldehyde (colorless crystals) in 50 g glacial acetic acid, 5 g *o*-phosphoric acid (85%) and 20 mℓ water. Solution can be kept for months if stored in a brown bottle.

Procedure: Spray plates and heat at 110°C for 10 min.

Comment: Cyclic sesquiterpene lactones give characteristic colors. Azulene hydrocarbons give blue spots.

	Daylight	UV (360 nm)
Gafrinin	bn-p	y
Geigerin	light y	
Vermeerin	light y	
Dihydrogriesenin	light bn	light y
Griesenin	bl	y

19. Dinitrobenzoic Acid

Preparation: Dissolve 1 g 3,5-dinitrobenzoic acid in a mixture of 50 mℓ methanol and 50 mℓ aqueous 2 *N* potassium hydroxide.

Procedure: Spray plates.

Comments: Certain cyclic sesquiterpene lactones and oxo compounds can be detected (geigerin-p).

20. Dinitrophenylhydrazine

Preparation: Dissolve 1 g 2,4-dinitrophenylhydrazine in 1 ℓ ethanol and 10 mℓ conc. hydrochloric acid.

Procedure: Spray plates.
Results: Aldehydes and ketones give yellow to red spots.
Comments: 0.3 g Sodium in 100 mℓ ethanol, overspray will intensify colors.

21. 2-Diphenyl-indane-1,3-dione-1-hydrazone
Preparation: Warm 0.5 g 2-diphenyl-indane-1,3-dione-1-hydrazine in 20 mℓ water, filter and add 0.3 mℓ 35% HCl.
Procedure: Spray plates and dry.
Results: Carotenoids and ketocarotenoids give blue to deep violet zones that fluoresce orange under UV light. Lower limit of sensitivity is 0.03 μg.

22. 2,2′-Diphenyl-1-picrylhydrazyl
Preparation: 15 mg 2,2-Diphenyl-1-picrylhydrazyl in 25 mℓ chloroform.
Procedure: Spray plates and heat at 110°C for 5 to 10 min.
Results: Terpenes give yellow spots on a purple background.

23. Emmerie-Engel Reagent
Preparation: (A) 2′ α,α′-dipyridyl in absolute ethanol; (B) 0.5% ethanolic ferric chloride.
Procedure: Spray with A then B.
Results: Ferric salt is reduced by tocopherol to ferrous salt, which forms a red complex with α,α′-dipyridyl.

24. Fast Blue
Preparation: Dissolve (0.1%) fast blue salt B in 1 *M* KOH. Prepare fresh.
Procedure: Spray chromatogram.
Results: Detects tetrahydrocannabinoids and other phenols.

25. Fluorescein
Preparation: 0.002% Sodium fluorescein in ethanol.
Procedure: Spray plates and examine under shortwave UV light.
Results: Detects menaquinones and aromatic and heterocyclic compounds.

26. Gibbs' Reagent
Preparation: (A) 0.05% *N*-2,6-Trichloro-*p*-benzoquinone in absolute ethanol (stable for 2 weeks in dark); (B) 4.75% sodium borate pH 9.2.
Procedure: Spray chromatogram with A then lightly with B.
Results: Detects polyprenylphenols with free *para* position. *Para*-substituted phenols generally do not react. Gives blue, purple, and red spots.
Comments: Overspraying with 10% acetic acid changes colors of phenolic compounds from blue to magenta; 4-hydroxy-3-methoxyl phenyl compounds become orange to yellow.

27. Godin's Reagent
Preparation: (A) 1% vanillin in ethanol; (B) 3% perchloric acid in ethanol.
Procedure: Spray chromatogram with A, then overspray with B. Heat at 80°C for 4 min.
Results: Detects oxygenated sesquiterpenes and diterpenes.

28. Hydrogen Chloride
Preparation: Concentrated HCl.
Procedure: Spray with conc. HCl.
Results: Carotenoid diepoxides give a rich blue color while monoepoxides turn blue green. Dihydroxy carotenoids (lutein and zeaxanthin) give a brown spot bordered with green. Other carotenoids give yellow to brown colors.

29. Iodine Vapor

Preparation: Place dried chromatogram into a closed tank containing iodine crystals.

Results: Most compounds will absorb iodine reversibly giving a brown spot on a yellow background. Especially useful for unsaturated compounds. Some compounds (chloramines, etc.) give white spots on a yellow background.

Comments: Color will fade after removal of the plate from the tank. The color can be preserved by placing a plate of glass over the chromatogram and taping the sides.

30. Liebermann Burchard Reagent

Preparation: 20% Conc. sulfuric acid in acetic anhydride. Add the acid to ice-cold anhydride. Caution: use in hood — highly lachrymatory.

Procedure: Spray chromatogram and detect under UV light.

Results: Detects sterols and triterpenols. Triterpenoid glycosides detected as red spots showing fluorescence under UV light.

31. Methanol-Sulfuric Acid

Preparation: Methanol-conc. sulfuric acid (1:1).

Procedure: Spray plates and heat at 110°C for 15 min. Inspect in daylight and UV light.

Results: Detects diterpene acetates.

	Color reaction (on silica gel) Daylight	UV (366 nm)
Phorbol triacetate	o	o
12-Deoxy-phorbol diacetate	r-bn	pk-bn
4-Deoxy-4α-phorbol triacetate	y-bn	y
4α-Phorbol triacetate	bl-gy	y-bn
4α-Phorbol tetraacetate	gy	y
Crotophorbolone monoacetate	pk	o
Ingenol triacetate	ol-bn	y

32. Neotetrazolium

Preparation: (A) 1% Sodium borohydride in ethanol-water (1:1); (B) 0.2% neotetrazolium in water. Dissolve the neotetrazolium in 4 mℓ of ethanol and bring to a final volume of 200 mℓ.

Procedure: Spray plates with A, and then overspray with B.

Results: Menaquinones appear as red spots without heating.

33. Orthophosphoric Acid

Preparation: Orthophosphoric acid - water (1:1)

Procedure: Spray plates and heat at 100°C. View under UV light and daylight.

Results: Detects sesquiterpene lactones.

	Color reactions Daylight	UV (360 nm)
Gafrinin	bn	w
Geigerin		bl-p
Geigerenin		w
Dihydrogriesenin		y
Greisenin	p	o-r

34. Perchloric Acid

Preparation: 70% Perchloric acid.

Procedure: Spray the chromatogram and heat for 5 to 10 min at 105°C.

Results: Naphthoquinones turn olive green or violet.

Comments: Detection limits are 0.4 to 4 μg.

35. Phosphomolybdic Acid

Preparation: 10 to 20% Phosphomolybdic acid in ethanol.

Procedure: Spray chromatogram and heat at 120°C for 10 min.

Results: A molybdenum blue color is obtained. Detects oxygenated monoterpenes, sesquiterpene lactones, alcohols, unsaturated hydrocarbons, and tocopherols.

Comments: Treatment with ammonia gas gives blue spots on a white background.

36. Phosphomolybdic Acid-2′,7′ Dichlorofluorescein

Preparation: Mix 1.6 g phosphomolybdic acid, 92 mg 2′,7′-dichlorofluorescein in 60 mℓ absolute ethyl alcohol, and 7.6 mℓ ammonium hydroxide. Dilute to 100 mℓ with water.

Procedure: Spray chromatograms until uniformly yellow. Inspect under daylight or UV (254 nm) light.

Results: Tocopherols give purple spots.

37. Potassium Permanganate

Preparation: 1% Aqueous potassium permanganate containing 6% sodium carbonate.

Procedure: Spray plates.

Results: Gives yellow spots on a reddish-violet background for phenolic compounds.

38. Potassium Permanganate-Copper Acetate

Preparation: 0.5% Potassium permanganate in saturated copper acetate.

Procedure: Spray chromatogram and inspect under daylight or heat to 110°C and examine under UV light.

Results: Oxygenated diterpenes give yellow colors if the chromatograms are not heated.

39. Rhodamine-B

Preparation: 100 mg Rhodamine B and 35 mg 2′,7′-dichlorofluorescein are dissolved in a mixture of 150 mℓ diethyl ether, 70 mℓ 95% ethanol, and 16 mℓ water.

Procedure: Spray chromatogram and allow to dry. Examine under UV light (366 nm).

Results: Lipids fluoresce orange or orange purple on a green background.

40. Rhodamine-6G

Preparation: 0.1% Aqueous Rhodamine-6G.

Procedure: Slurry silica gel with the above solution (55 mℓ for 25 g silica gel G) when preparing layer.

Results: Detects sterols under UV light as pink-yellow fluorescent spots on pale background. UV-absorbing compounds appear as dark spots.

41. Rhodamine-Ammonium Hydroxide

Preparation: (A) Rhodamine-B or 6G (1 to 5%) in ethanol; (B) 25% ammonium hydroxide.

Procedure: Spray with A and B and dry.

Results: Carotenals give deep violet zones with a lower limit of sensitivity of 0.03 μg.

42. Rosenberg's Reagent

Preparation: (A) 3 mℓ 5% Ammonium molybdate and 7 mℓ 5 *N* hydrochloric acid are brought to 100 mℓ with acetone. (B) 2 g vanadium pentoxide is added to 20 mℓ of boiling conc. hydrochloric acid. Continue boiling for 10 min after the color of the solution turns blue-green. Use 6 *N* hydrochloric acid to 400 mℓ with water (0.5% V_2O_5 in 0.5 *N* hydrochloric acid). Bring 20 mℓ of the vanadyl chloride to 100 mℓ with acetone. Add 250 mg fine zinc powder. Shake until clear and brown in color. Decant.

Procedure: Spray chromatogram with A, dry, and dip in B.

Results: Detects phosphorylated isoprenoids.

43. Sodium Methoxide-Dimethylsulfoxide

Preparation: Saturated solution of Na metal (8 g) in methanol-dimethyl sulfoxide (100:8).

Procedure: Spray and while still wet, examine under long wave UV light.

Results: Tetrahydrocannabinoids give a fluorescent yellow-green spot.

44. Sulfuric Acid

Preparation: Conc. sulfuric acid or 60% sulfuric acid.

Procedure: Spray and heat to 100°C for 15 min.

Results: Charring is often preceded by specific colors or fluorescent spots. A general reagent.

45. Toluenesulfonic Acid

Preparation: 20 g *p*-Toluenesulfonic acid in 100 mℓ ethanol.

Procedure: Spray chromatogram and heat at 100°C for 3 to 5 min. Inspect under filtered UV light.

Results: Detects sesquiterpene lactones and sterols.

	Color reactions Daylight	UV (360 nm)
Gafrinin	bn-p	o-y
Geigerin	light y	bl
Geigerinin		light y
Dihydrogriesenin	y	b
Griesnin	bk-p	y-bn

46. Trichloroacetic Acid

Preparation: 25 g Trichloroacetic acid in 100 mℓ chloroform.

Procedure: Spray chromatogram and heat at 100°C for 2 min.

Results: Detects sesquiterpene lactones (yellow colors) and sterols.

47. Vanillin-Sulfuric Acid

Preparation: 1 to 5% Vanillin in conc. sulfuric acid.

Procedure: Spray plate and heat at 110° for 5 min.

Results: Detects oxygenated monoterpenes, sesquiterpenes, and diterpenes.

Color reactions (on silica gel)

Monoterpenes	
Menthone	gy
Dihydrocarvone	rbn
Pulegone	gy-bl
Terpineol	o,gn
Carvone	o-r
Piperitone	y-o
Carvenone	y-gr
Piperitenone	y-o
Limonene	o,gn
Geranyl acetate	p,vt
Linalyl acetate	p,vt
Linalool	o,vt
Citral	bn

Diterpenes	**Daylight**	**UV (366 nm)**
Phorbol triacetate	p-bn	bl
12-Deoxy-phorbol diacetate	gy-bn	r
4-Deoxy-4α-phorbol triacetate	p-bl	bl
4α-Phorbol triacetate	bl	dark bl
4α-Phorbol tetraacetate	bl	dark bl
Crotophorbolone monoacetate	p	bl
Ingenol triacetate	gy-bn	y

48. Vanillin-Ethanol-Sulfuric Acid

Preparation: (A) 1% Vanillin in ethanol; (B) 60% aqueous sulfuric acid.
Procedure: Spray plates with A, then overspray with B. Heat at 110°C for 5 min.
Results: Detects diterpenes.

Section III
Sample Preparation and Specific Detection Techniques

SAMPLE PREPARATIONS

A. MONOTERPENES

1. General Aspects

Because of their low molecular weight and hydrocarbon-like structures, most common monoterpenes are liquids that are readily separated by GC without derivatization. Capillary columns have been especially effective, exhibiting high resolving capacity. While flame ionization has provided a high sensitivity for detection of column effluents, mass spectrometry has been far more definitive for essential oil identification[1,2] (Figures 1,2,3). On-line computers interfaced with the mass spectrometer can assess less intense ion peaks, thereby contributing to the reliability of the assignments.[3]

HPLC offers great promise for the resolution of isomeric monoterpenes (see Table LC 1), and for compounds with chromophores, a UV detector affords sensitivity in the nanomole range.[3a] Semipreparative HPLC may be useful in sample preparation of monoterpene mixtures (Figures 4 and 5).

2. Sample Preparation from Plant Material

a. Essential Oils by Steam Distillation

A simple steam distillation protocol was reported by Kumamoto et al.[6] Leaf tissue was homogenized with 1.4 volumes of water in a carbon dioxide atmosphere in a Waring blender. Steam distillation was then performed with a Clevenger apparatus for oils less dense than water. The cold finger of the apparatus was maintained at 0°C with a refrigerated cooling solution to ensure recovery of more volatile terpenes. The distillate was stored frozen until used.

In some instances the steam distillate is extracted with ether or some other appropriate organic solvent. Micro-steam distillation may be used to avoid large losses due to evaporation and terpene solubility in ether.[7]

b. Headspace Volatiles

Aroma profiles may be obtained by direct GC analysis of headspace components. Routine vapor sampling was achieved by placing the plant material (1 g) into a microvial fitted with an on-off valve and septum. After equilibration for 1 hr (at 75°C, 5 mℓ of the headspace air containing the volatiles was withdrawn using a gas-tight syringe and immediately injected into a gas chromatograph. Aroma profiles included compounds present in the headspace at concentrations as low as 0.1 ng/mℓ, which is below the human odor threshold for many odorants.[8]

c. Direct Injection

A rapid, simple method for essential oil analyses employed an inductor for direct injection of plant material into a pyro-oven at 160 to 250°C for short time intervals to prevent pyrolysis of the residue.[9]

d. Monoterpene Glycosides

For water-soluble monoterpenes such as the iridoid and secoiridoid glycosides, water or methanol is used to extract the plant material and the extract is then subjected to an initial fractionation by column chromatography to separate simple carbohydrates.

Various silica gel preparations were useful for separation of radioactive monoterpene glucosides from extracts of *Swertia caroliniensis* by the following procedure[10]: After homogenization of the plant material (9 g) in methanol (900 mℓ) with a Waring blender, the solvent was filtered from the mash and evaporated. The crude methanolic extract (1.5 g)

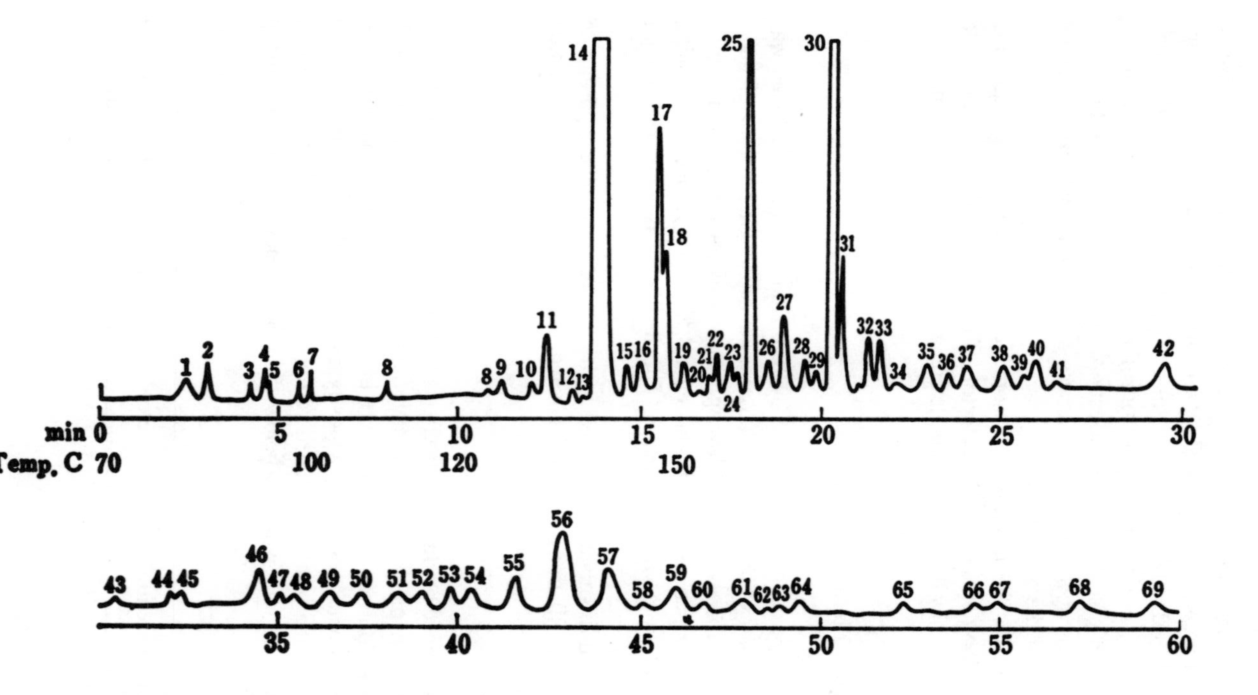

FIGURE 1. Gas chromatogram of the constituents of an essential oil.

Instrument: Hitachi Gas Chromatograph K-53.

Column: 45 m × 0.25 mm I.D. stainless steel capillary column coated with polypropylene glycol-2000.

Detector: Either FID or rapid scan mass spectrometry (Hitachi Mass Spectrometer RMU-6 ionization voltage 80 eV, ion source temperature 250°C, ion accelerating voltage 2 kV, and inlet vapor temperature 150°C).

Carrier gas: Helium.

Temperature: 70-150°C at 5°C/min.

Compounds: See Table III-2 for compound assignments.[4]

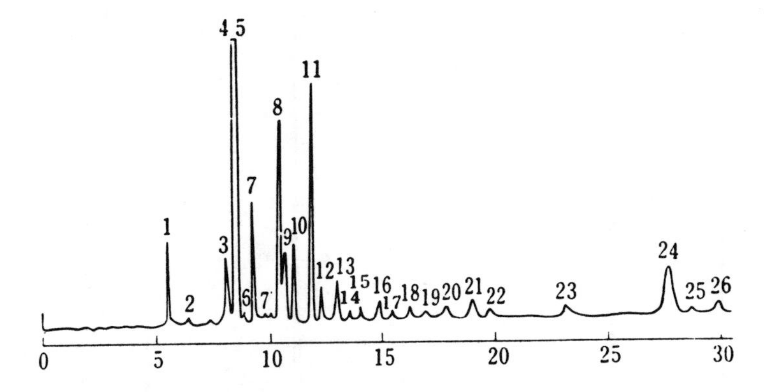

FIGURE 2. Gas chromatogram of the sesquiterpenic hydrocarbon fraction of an essential oil. Conditions as in Figure 1 except for the column temperature which was maintained at 150°C. See Table III-2 for compound assignments.

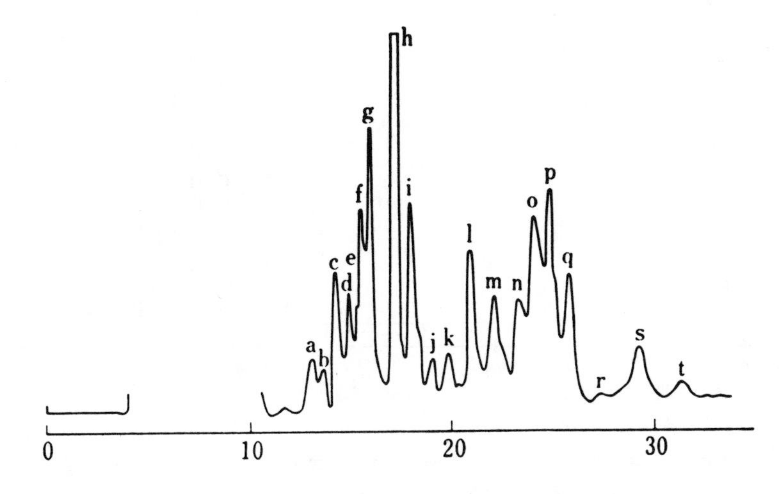

FIGURE 3. Gas chromatogram of the sesquiterpenic alcohol fraction of an essential oil. Conditions as in Figure 2. See Table III-2 for compound assignments.

was chromatographed on a silica gel G column (1.5 × 55 cm), eluting with 500 mℓ each of chloroform-methanol (4:1 v/v), chloroform-methanol (1:1 v/v), and chloroform-methanol-water (4.5:4.5:1 v/v) under 2.5 psi nitrogen pressure. The first eluent afforded gentiopicroside, which was characterized as its tetraacetate, whereas loganic acid occurred in 25 mℓ fractions 31 to 34 of the last solvent mixture used. The latter (120 mg) was further purified on two successive silica gel H columns (1.5 × 55 cm), eluting with chloroform-methanol (2:1 v/v). This silica gel H was prewashed with formic acid and dried at 110°C. Loganic acid fractions that did not contain traces of front-running or trailing impurities were combined and chromatographed on a silica gel column (1.5 × 30 cm, Merck silica gel, particle size less than 0.08 mm, prewashed with formic acid and dried), eluting with chloroform-methanol (1:1 v/v). The loganic acid fractions were next subjected to preparative thin layer chromatography with a developing solvent of chloroform-methanol-water (5:5:0.8 v/v) on silica gel H plates (0.75 mm thickness; the plated silica gel was "prewashed" by allowing ether saturated with concentrated formic acid to run up the plate three times). This was repeated twice with no change in specific activity of loganic acid. The glycoside (35 mg) was finally chromatographed on a prewashed silica gel H column (1.5 × 30 cm) eluted with chloroform-

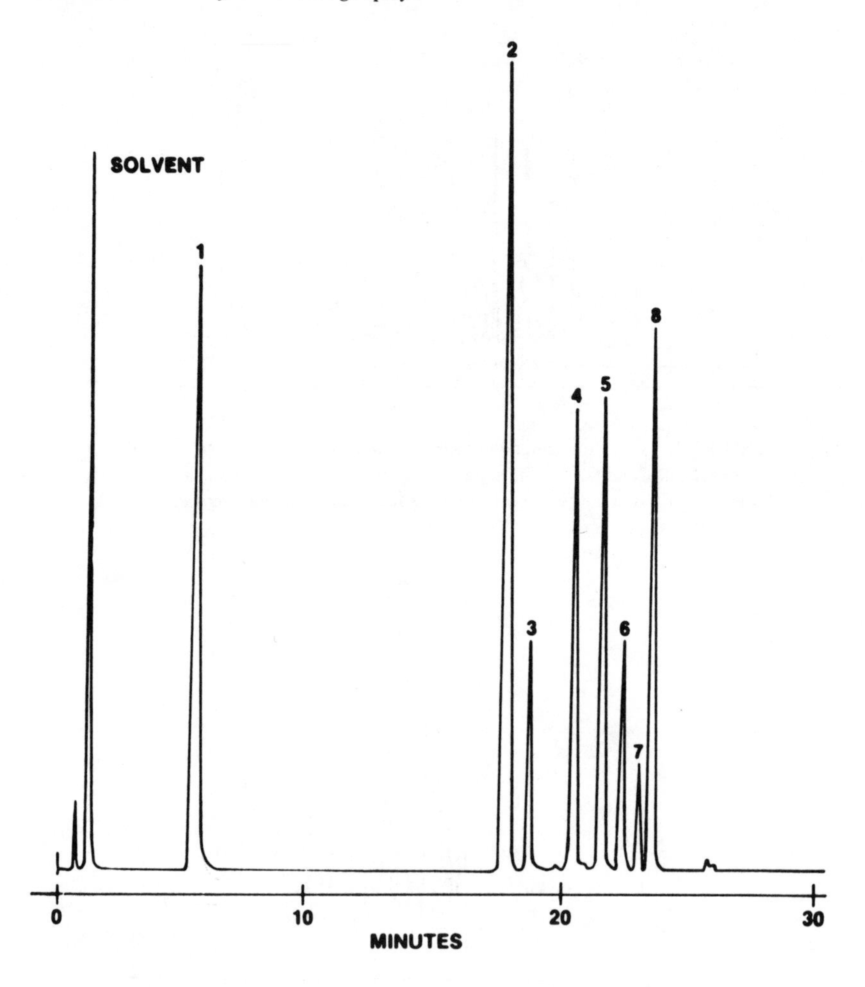

FIGURE 4. Analytical GC separation of a model mixture of monoterpenes.
Column: 12 ft glass column containing 5% Triton X-305.
Temperature: 70-170°C at 5°C/min, 5 min initial time.
Compounds: 1=limonene, 20%; 2=*trans-p*-mentha-2,8-dien-1-ol, 24%; 3=*cis-p*-mentha-2,8-dien-1-ol, 6%; 4=carvone, 14%; 5=*trans-p*-mentha-1(7),8-dien-2-ol, 12%; 6=*trans*-carveol, 6%; 7=*cis*-carveol, 3%; 8=*cis-p*-mentha-1(7),8-dien-2- ol, 14%.[5]

methanol (2:1 v/v), with no change in specific activity. The value obtained agreed well with that of the loganic acid isolated from the excised leaf that had been diluted (42-fold), acetylated, methylated, and recrystallized to constant specific activity.[10]

Carboxylic acid derivatives of water soluble glycosides can be resolved from neutral monosaccharides and glycosides readily as follows: Methanolic extracts of *Catharanthus roseus* were evaporated to dryness, and the residue was washed with hexane. The extract was redissolved in water and subjected to ion-exchange column chromatography on Dowex 1-X8 (200-400 mesh), formate form (200 × the theoretical milliequivalents). After washing the column with water to remove sugars and other neutral substances, elution with 0.1 *N* formic acid was initiated. Fractions 1 to 3 contained unidentified carboxylic acids. Secologanic acid was the major component of fractions 4 to 6, giving an R_F 0.6 on thin layer plates coated with silica gel GF_{254} and developed with ethyl acetate-methanol-formic acid (4:1:0.5). Loganic acid was eluted in fractions 8 and 9. Further purification of fractions 4 to 6 by ion-exchange chromatography with linear gradient elution (H_2O - 0.1 *N* formic acid) led to isolation of secologanic acid in pure form, which upon evaporation under reduced pressure formed a foamy white residue.[11]

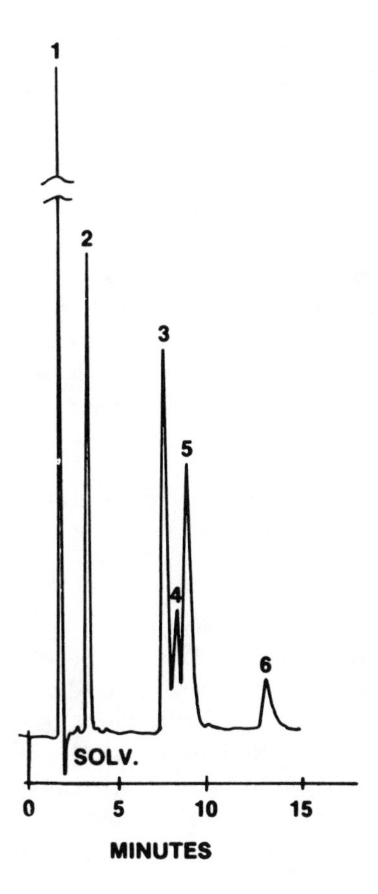

FIGURE 5. Semipreparative HPLC separation of a model mixture of monoterpenes.

Instrument: Waters ALC/GPC 201 HPLC system including a M-6000 pump, a M-U6K universal injector and a M-R401 differential refractometer.

Column: Whatman prepacked Partisil 10-PXS (10 μm silica gel) 25 cm × 4.6 mm I.D., stainless steel.

Solvent, flow rate: Ethyl acetate-hexane (10:90), 2 mℓ/min.

Compounds: 1=limonene; 2=carvone; 3=*trans*- and *cis*-*p*-mentha-1(7),8-dien-2-ol; 4=*trans*- and *cis*-carveol; 5=*trans*-*p*-mentha-2,8-dien-1-ol; 6=*cis*-*p*-mentha-2,8-dien-1-ol.[5]

After the above prefractionation to separate nonpolar derivatives, subsequent conversion of polar metabolites to hydrophobic derivatives facilitates their resolution. The acidic components from *C. roseus* plants (1 kg), obtained by ion-exchange chromatography as described above, were subjected to diazo-methylation. Most of the loganin obtained from loganic acid was recrystallized from acetone, and the mother liquors were subjected to silica gel G column chromatography, the eluting solvent being chloroform-methanol (9:1). Fractions less polar than loganin were pooled and acetylated with an equal volume of acetic anhydride and pyridine. After the usual work-up, the residue was purified on preparative silica gel GF_{254} plates developed with benzene-hexane-methanol (4.5:4.5:1). The band having an R_F of 0.48 was rechromatographed with benzene-ethyl acetate (2:1) as the developing solvent. The major band was extracted from the silica gel with acetone to afford, after evaporation to dryness, 60 mg of residue that was recrystallized from ethanol to give a homogeneous compound, secologanoside methyl ester tetraacetate.[11]

3. Sample Preparation from Urine

Terpenes fed to animals in large quantities are metabolized usually by hydroxylation and oxidation. Often hydration of double bonds is followed by glucuronidation of hydroxyl groups.

a. Simple Extraction of Limonene Metabolites

Kodama and his colleagues[12] have developed a systematic prefractionation of neutral and acidic limonene metabolites. Urine samples collected 48 hr following administation of d-limonene were adjusted to pH 2.0 with conc. HCl and continuously extracted with ethyl acetate for 40 hr. The aqueous phase was retained for analysis of hydrophilic metabolites, whereas the organic phase was re-extracted with 1.5 *N* aqueous NaOH. Neutral metabolites remained in the ethyl acetate and acidic compounds were recovered from the alkaline aqueous phase by adjustment of the pH to 2 and sequential extraction with petroleum ether, benzene, ethanol, and ethyl acetate.

The original extracted urine was passed through an Amberlite XAD-2 (Rohm and Haas, Philadelphia, Pa. U.S.A.) resin column (5 × 45 cm). A methanol eluate contained a glucuronide derivative.[12]

b. Simple Extraction of Camphor Metabolites

Pretreatment of urine with β-glucuronidase will convert monoterpene conjugates to their free form, but unconjugated metabolites are removed first by extracting 24 hr and 48 hr urine samples 6 times with equal volumes of ether. Residual ether was removed with a rotary evaporator, and the urine was brought to pH 6.8 to 7.0 with 0.1 *M* Sörensen's phosphate. Incubations with β-glucuronidase (bacterial type II; Sigma Chemical Co., St. Louis, Mo., U.S.A.) and a few drops of chloroform were performed at 37°C for 24 hr. The mixture was cooled and extracted with ether to obtain hydrolyzed monoterpenes.[13]

4. Hydrogenation and Hydrogenolysis of Monoterpenes

Carbon skeleton analysis of small amounts of monoterpenes has been accomplished by hydrogenation and hydrogenolysis followed by mass spectral detection of gas chromatographic effluents. The method is useful in detecting classes of terpenes rather than individual compounds, but suffers from variability in catalyst activity.

To prepare 2% palladium catalyst on acid-washed Chromosorb W, 60-80 mesh, palladium chloride (330 mg) was dissolved in 100 mℓ of 5% aqueous acetic acid. This was accomplished by heating for 1 hr with occasional swirling. Upon cooling and addition of anhydrous sodium carbonate (200 mg), the solution was evaporated to dryness in the presence of Chromosorb W (10 g) in a rotary film evaporator. The catalyst was subjected to oven heating at 110°C for 1 hr. The authors report that only one in every three preparations was sufficiently active to catalyze complete hydrogenation.

Activation of the catalyst for hydrogenolysis was accomplished by packing approximately 0.5 g of the catalyst into the injection port tube (12 × 0.5 cm) to afford a 10 cm column. Without being attached to the GC column, the injection tube was flushed with hydrogen at a flow rate of 25 mℓ/min and heated for 30 min at 150°C and 33 min at 280°C. The oven remained open during this period. The hydrogenolysis potency of the catalyst normally lasted about 10 days.

The chromatograph used was a Model 5750 Hewlett Packard instrument equipped with dual flame ionization detectors and a Carle thermistor detector. The hydrocarbons were separated on 10 ft × 1/8 in. O.D. stainless steel columns packed with 5% Apiezon L on 60-80 mesh Chromosorb W, A.W., or 500 ft × 0.03 in. I.D. stainless steel capillary columns coated with Carbowax 20M or SF-96.[50] To prevent overheating of the septa in the injection tube during hydrogenolysis, the exterior ends of the injection port were fitted with water-cooled jackets.

In hydrogenolysis experiments, the 10 ft Apiezon L column and thermistor detector were utilized. The injection port, column, and detector temperatures were maintained at 275, 90, and 180°C, respectively. Hydrogen at a flow rate of 25 mℓ/min was the carrier gas. Mixtures were injected in *n*-pentane, and reaction products were collected in 12 in. melting point capillaries cooled in liquid nitrogen. Combined GC-MS was performed with a Varian MAT CH5 mass spectrometer. In this case, either the Carbowax 20M or the SF-96[50] capillary columns was used. Infrared spectra were obtained with a Perkin Elmer Model 257 spectrophotometer. Samples were dissolved in carbon tetrachloride and placed into a ultra-micro cavity cell. A variable path length cell containing carbon tetrachloride was placed in the reference beam.

Under the conditions used, acyclic monoterpenes gave the parent saturated hydrocarbon with the exception of primary alcohols, which afforded a major product having lost the CH_2OH group. Cyclopropane rings were cleaved, four- and five-membered rings remained intact, and cyclohexyl groups gave predominantly aromatic products with small amounts of isomeric menthanes.[14]

5. Monoterpene Alcohol Carbamates

In complicated essential oil mixtures tertiary alcohols can be difficult to separate, and when subjected to gas chromatography they undergo dehydration. By the reaction of trichloroacetyl isocyanate with the alcohols and then hydrolysis of the resultant esters to carbamates with aqueous methanolic KOH, the alcohols can be isolated from other constituents in the essential oil.

A 30 mℓ solution of about 0.5 *M* alcohol in carbon tetrachloride was treated with a 30% molar excess of trichloroacetyl isocyanate (0.023 mol in 2.2 mℓ), and the mixture was allowed to stand in an ice bath for 5 to 10 min. It was then extracted with 50 mℓ of 5% KOH in methanol-water (1:4). The aqueous basic methanol phase was removed and evaporated *in vacuo* to an aqueous residue that was then extracted with CCl_4 until TLC revealed all carbamates had been extracted. The crude carbamates could be resolved by TLC or GC[15] (see Table TLC 3).

B. SESQUITERPENES

1. General Aspects

Since sesquiterpenes are at times structurally similar to monoterpenes, sample preparations for certain groups of monoterpenes can be applied to the related sesquiterpenes. However, steam distillation methods need to be more vigorous to obtain the oxygenated components.

In fact, one can predict sesquiterpene GC retention times by correlating them with values of monoterpenes that differ structurally by only a single isoprene unit.[16] A semilog plot of relative retention time vs. number of carbons is constructed, and the change, as indicated by the slope, is determined with standards. For the reference compounds used, the slopes were constant.

Since sesquiterpenes have higher molecular weights than the C_{10} compounds, it is at times desirable to prepare volatile derivatives of their alcoholic forms before subjecting them to gas chromatography. Sesquiterpenes that are not too highly oxygenated can be prefractionated by steam distillation (see Section III.A.2.a).

2. Sample Preparation of Sesquiterpenoid Mold Metabolites

Sesquiterpenes are produced by microorganisms that are found in soil and intact plants. The trichothecene group of mycotoxins represent one such group. The following represents a typical preparation.

Shelled corn was kept in an equal amount of tap water overnight at room temperature and then sterilized (121°C, 15 psi, 1 hr). Upon cooling a suspension of *Fusarium tricinctum* was added and after 48 hr at room temperature, the culture was maintained at 8°C for 30 days. Lyophilization was followed by grinding of the culture in a Wiley mill and storage at −20°C.

The fine powder was extracted with two 25-volume portions of ethyl acetate in a Waring blender. The ethyl acetate extract was then washed with water three times and the organic phase concentrated to a thick brown oil *in vacuo*. The latter was dissolved in about $^1/_3$ of the original volume of methanol-water (4:1), which was then extracted three times with Skellysolve B. The methanol phase was diluted with water to a final ratio of 1:1 and extracted with three equal volumes of chloroform-ethyl acetate (1:1). The organic layer was taken to dryness *in vacuo*, and the residual light oil was silylated for GC with bis-(trimethylsilyl)acetamide reagent. D-Glucose was used as an internal standard and chromatographed as described in Table GC 17. Corn-rice mixtures were treated with the same fungi, but incubations were at 27°C for 21 days and 8°C for 24 days. If the mixture cannot be gas chromatographed directly because of interfering substances, prefractionation on silica gel is included.[17]

3. Hydrogenation and Hydrogenolysis

Maarse has subjected acyclic, mono-, bi-, and tricyclic sesquiterpenes to the injection port hydrogenation method originally developed by Beroza (see Table GC 14). As with his studies with monoterpenes, this procedure is valuable when only small amounts of sesquiterpenes are available. Several acyclic and monocyclic sesquiterpenes gave expected products, but ten-membered monocyclic compounds such as germacrenes and germacrones give a large number of products that are mostly bicyclic. While the bicyclic sesquiterpenes, selinenes, gave two major products, cadinanes, murolanes, and amorphanes formed more compounds than anticipated. Tricyclic compounds, thujopsene, α-cedrene, and longifolene gave the two types of products obtained by hydrogenation in solution.[18]

4. Dehydrogenation

Sesquiterpenes can be dehydrogenated to naphthalenes and azulenes providing products that give highly reproducible gas chromatographic retention data with even small amounts of starting material. In some cases characteristic patterns may be obtained for individual compounds. The following procedures are for 250 mg essential oil samples. Dehydrogenation using elemental sulfur (0.1 g) were performed by refluxing in 3 to 10 volumes of triglyme for 2 hr. Light petroleum ether was added to the reaction mixtures, and the solution was extracted repeatedly with water, 85% aqueous H_3PO_4, 10% aqueous NaOH, and saturated sodium chloride. After drying the organic phase over Na_2SO_4, it was passed over alumina (Woelm, basic, activity I-II) before GC or NMR spectroscopy. Azulenes were recovered from the phosphoric acid layers with petroleum ether after addition of water.

When selenium was used for dehydrogenation, the sesquiterpenes (neat) were treated with an excess (0.5 to 1.0 g) at 260 to 320°C for 30 min. The petroleum ether extract was prefractionated on alumina (Woelm, basic, activity I) before GC or NMR spectroscopy. Fractions containing the blue azulenes were purified by repeated extractions using petroleum ether and phosphoric acid as mentioned above.[19]

C. DITERPENES

1. General Aspects

In general, diterpenes are relatively hydrophobic molecules resembling triterpenes in their chromatographic behavior. An important group are the gibberellins, a family of plant hormones that regulate growth and differentiation of higher plants. They contain five rings and are highly oxygenated. As many as 52 free isomeric gibberellins have been characterized.[20]

2. Extraction of Gibberellins from the Fungus *Gibberella fujikoroi*

Gibberellins are produced in small amounts by most plants, but the fungus *Gibberella fujikoroi* is an excellent source. By varying conditions, cultures normally containing gibberellic acid (A_3) can form A_4 or A_7 as well as other related isomers. The following procedure has been described by Pitel et al.[22] and is quoted here.*

The mycelium was removed by filtration. The filtrate was acidified to pH 2 to 2.5 and, if the volume was not too large, extracted directly with ethyl acetate (3 × $^1/_2$ volume). Extraction of the ethyl acetate solution with 0.5 *M* sodium carbonate gave an acidic fraction that was recovered and evaporated. Large volumes of filtrate (i.e., over 2 ℓ) were stirred for 5 hr with activated charcoal (The British Drug Houses Ltd., Montreal, Quebec). The washed charcoal cake, dried to approximately 40% water content, was extracted thoroughly with acetone. Concentration of the extract gave an aqueous solution that was extracted with ethyl acetate as above.

For preliminary fractionation of acidic metabolites a column of silicic acid was used. Two procedures were tested and found suitable. In the first, silicic acid (No. 2847, 100 mesh size; Mallinckrodt Chemical Works) treated with 0.5 *M* formic acid was developed with a gradient of ethyl acetate in petroleum ether, as described by Powell and Tautvydas.[21] In the second, silicic acid (SilicAR CC4, Mallinckrodt Chemical Works) was irrigated with benzene, chloroform, ethyl acetate, and methanol in a stepped eluotropic series.

The column was prepared was follows. The cross-linked dextran (Sephadex G-25 in fine bead form, purchased from Pharmacia Ltd., Montreal, Quebec) was allowed to swell for 1 to 2 hr in the aqueous phase of a biphasic solvent mixture. The gel was collected in a large sintered glass funnel, and excess solvent drawn off by suction. It was then suspended in the organic phase and the slurry added, in portions, to a 100 cm long glass column filled to 1/3 volume with organic phase. The gel was dispersed evenly in the column by a few rapid strokes of a packing tool (a stainless steel rod terminating in a perforated disc matching the column diameter). The stopcock on the column was then opened, and the gel compressed by means of the packing rod.

With some solvent mixtures, particularly those containing butanol as the water-immiscible component, satisfactory columns can be prepared by gravity settling of the gel. However, for mixtures in which the aqueous phase is lighter than the organic phase, use of a packing tool is essential. Columns were of several diameters, the largest being 2.6 cm. However, no upper limit was established, and the selection of column size depends mainly upon the size of sample and its solubility in the stationary phase. Solvents were mixed in the proportions specified by volume. The gel was equilibrated before use by washing with organic phase for 18 hr, or until no aqueous phase was present in the effluent. After use, the column could be washed with aqueous phase and reequilibrated with organic phase without repacking.

The sample was dissolved in the minimum volume of aqueous phase and the solution added to sufficiently dry Sephadex G-25 to saturate the gel. For this calculation a solvent regain value of 2.5 g/g (the water regain quoted by the manufacturer) was used. The gel sample was then packed on the top of the Sephadex column and covered with a filter paper disc. The column was developed with organic phase of the solvent system at a flow rate of approximately 1 mℓ/min for a 2.6 cm diameter column, and the eluate collected in 20 mℓ fractions.

3. Extraction of Conjugated Gibberellins

Twelve conjugated gibberellins have been detected in seedlings, immature and/or mature seeds. These glucosides were subjected to solvent extraction, charcoal chromatography, partition chromatography, and preparative thin layer chromatography (PTLC) before HPLC.

Immature seeds were homogenized in a Waring blender with methanol. After three extractions the methanol solutions were pooled and concentrated *in vacuo* to yield a residual aqueous solution. The solution was brought to pH 7 and extracted three times with benzene and then ethyl acetate. After acidifying to pH 3 the aqueous solution was extracted again with ethyl acetate and *n*-butanol. The *n*-butanol fraction showed high gibberellin activity in a bioassay using rice seedlings.

The *n*-butanol fraction was then subjected to charcoal column chromatography using a step gradient of acetone-water. The acetone was increased in increments of 10% for each 500 mℓ of eluent. The 40, 50, and 60% aqueous acetone eluates were enriched in gibberellins. The 40 and 50% aqueous acetone eluates were combined and rechromatographed on charcoal, eluting with acetone-water. Again 10% steps of acetone were used for each 300 mℓ of eluent. Fractions eluted with 6% aqueous acetone from the first column and 50 and 60%

* **Pitel, D. W., Vining, L. C., and Arsenault, G. P.,** *Can. J. Biochem.*, 49, 185, 1971. With permission.

aqueous acetone from the second column were combined evaporated to dryness to afford a gum. The latter was subjected to partition chromatography using Sephadex G-50 impregnated with 1 *M* phosphate buffer (pH 5.5). The eluent was ethyl acetate-*n*-butanol (9:1) and the fractions were monitored by TLC using silica gel plates developed with chloroform-methanol-acetic acid (50:20:3). The eluate obtained after approximately 100 column volumes showed the characteristic fluorescence at 3650 Å on TLC plates after spraying with 70% sulfuric acid and heating. Gibberellin-containing fractions were combined and subjected to preparative TLC on silica gel with the same solvent system mentioned above. The band migrating with an R_F from 0.2 to 0.4 was scraped off the plates and extracted with methanol. The methane extract was concentrated and fractionated on a Sephadex LH-20 column eluted with *n*-butanol-water (6:100). Fractions giving the characteristic fluorescence described above were then subjected to HPLC analyses.[20]

4. Preparative Scale Chromatographic Separation of Gibberellins

Another method of isolating gibberellins and related diterpenes entails elution of mixtures from Sephadex LH-20 columns. The eluent was a biphasic solvent system made from light petroleum, methanol, and acetic acid to which ethyl acetate and water was added to give two phases.[23]

D. TRITERPENES

1. General Aspects

Triterpenes, like steroids, are often isolated from a neutral fraction after saponification of lipid mixtures, and are therefore referred to as nonsaponifiable. Like other higher terpenoids the introduction of oxygen in triterpene molecules reduces their volatility considerably. Frequently analysis by GC is performed on trimethylsilyl or fluoroacyl derivatives. Due to their relatively low volatility, complex triterpene mixtures have not been directly resolved by GC. Structural diversity arising from stereoisomerism obviates fractionation by TLC and only recently have they been resolved with the aid of HPLC (Figure 6).

2. Sample Preparation

a. Plant Material

Plant material is usually extracted with light petroleum or chloroform to give an oil that is in some cases saponified. Alternatively prefractionation on alumina or silica gel columns or TLC plates can be performed before a final resolution step.

b. Saponification of Vegetable Oils

Vegetable oils have been saponified in 10 volumes of 10% alcoholic potassium hydroxide by refluxing for 1 hr under nitrogen. The reaction mixture was diluted with two volumes of water, and the unsaponifiables were extracted with isopropyl ether ($^1/_3$ volume once, three times with a $^1/_4$ volume). After combination, the organic phase was washed with 1/5 volume of water several times, and dried over anhydrous sodium sulfate. Prefractionation on preparative silica gel layers or on alumina or silica gel columns was performed before gas chromatographic analyses. Plant extracts may be treated in a similar fashion after air drying, maceration, and extraction.[25]

c. Acid Hydrolysis of Sterol Glycosides

Dioscorea tubers contain a family of sterol glycosides that are an important starting material for commercial steroid syntheses.

After a maceration of the tubers, they were dried at 100°C under atmospheric pressure and pulverized in a blender for several minutes. A 0.6 g sample and 15 mg of an internal

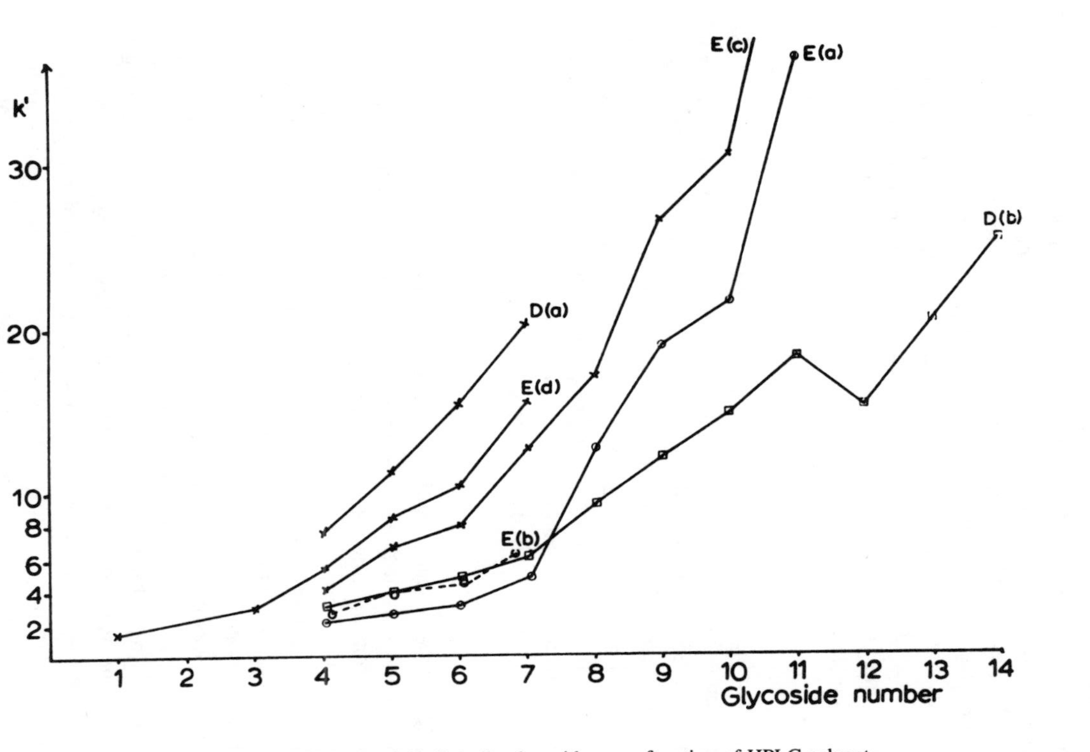

FIGURE 6. Capacity coefficients of 14 digitalis glycosides as a function of HPLC solvent.

Instrument: Hewlett-Packard Model 1010A equipped with a UV detector monochromator HP Model DSF 1000.

Columns: LiChrosorb SI-60 of particle size 10 μm in stainless steel columns of 25 cm × 3.0 mm I.D.

Solvent, flow rate: D (a)=t-butanol-acetonitrile-heptane-water (22:7:80:10), 2.2 mℓ/min; E=*n*-pentanol-acetonitrile-iso-octane-water (a) 270:93:660:9.3; (b)23:10:70:10; (c) 175:60:620:10; (d) 175:60:620:6; 1.3-2.5 mℓ/min.

Compounds: Glycoside number 1=digitoxigenin, 2=gitoxigenin; 3=digoxigenin; 4=digitoxin; 5=gitoxin; 6=digoxin; 7=diginatin; 8=lanatoside A; 9=lanatoside B; 10=lanatoside C; 11=lanatoside D; 12=deacetyl-lanatoside A; 13=deacetyl-lanatoside B; 14=deacetyl-lanatoside C.[24]

standard were placed in a hydrolysis tube. If the sample was not completely dry, it could be dried to constant mass by heating overnight at 100°C *in vacuo* (30 mmHg). To the dry sample was added 8 mℓ of xylene and 8 mℓ of 3 *N* HCl and the mixture was refluxed for 4 hr on a hot plate with stirring. After initial foaming, the mixture was refluxed vigorously at 93°C. On cooling the clear upper layer of organic phase was removed and subjected to GC analyses.[26]

3. Benzoate Derivatization

With the exception of a few subclasses, triterpenes do not have chromophores absorbing in the UV or visible range. In fact even cardienolides (digitalis glycosides) exhibit only a weak absorbance with a λ_{max} at around 220 nm. Quantitation of complex mixtures has been difficult and depends on relatively inaccurate colorimetric reactions, which, if used after resolution by TLC, can require tedious densitometric techniques. On the advent of the HPLC of triterpenes, it became clear that efficient and simple derivatization of these compounds with aromatic reagents would facilitate their detection and quantitation. To this end several approaches have been taken (see Tables LC 7 and 9).

a. Benzoates

A weighed sample containing approximately 25 mg of sapogenin was dissolved in 2 mℓ of pyridine in a 40 mℓ screw cap test tube with a PTFE-liner. After 0.1 mℓ of benzoyl chloride (reagent grade) was added, the sample was heated in a water bath at 80°C for 30 min. After cooling, 10 mℓ of redistilled dichloromethane, 10 mℓ of distilled water, and 2.0 mℓ of conc. HCl were added in that order. The tube was capped, shaken for 15 sec, and the aqueous layer aspirated off. The washing was repeated with another 10 mℓ of water.

The precision of the benzoylation procedure was checked, and results were found to have a 95% confidence interval of ±1.56%. Benzoate esters were monitored at 235 nm.[28]

b. 4-Nitrobenzoates

The reagent was freshly prepared daily by dissolving 100 mg of 4-nitrobenzoylchloride in 1 mℓ of pyridine with gentle warming. The acid chloride (analytical grade) was purified by a single recrystallization from light petroleum ether (b.p. 60 to 70°C).

To 50 μℓ of a pyridine solution containing not more than 0.5 mg of the glyceride in a stoppered 10 mℓ centrifuge tube was added 150 μℓ of the reagent solution. The mixture was shaken vigorously and kept for 10 min at room temperature after which time the reaction was found by TLC to be quantitative. the centrifuge tube was immersed in a sand bath that was heated to 50°C. The bath was placed in a desiccator and the solvent evaporated off in 10 min by connecting the desiccator to a water aspirator. After flushing the tube with a stream of air or nitrogen, a 2 mℓ solution of 5% aqueous sodium carbonate containing 5 mg of 4-dimethylaminopyridine was added. The latter hydrolyzed excess reagent after 5 min of shaking or sonication. The esters give turbid solutions due to their insolubility in water, while a blank yields clear solutions. To remove reagent and pyridine the aqueous solution was extracted with 2 mℓ chloroform that was then washed once with 2 mℓ of 5% aqueous sodium bicarbonate and twice with 3 mℓ of 0.05 *N* hydrochloric acid containing 5% of sodium chloride.

The derivatives were monitored at 260 nm. Since they possess high absorbance, detection limits were below 20 ng/mℓ if injections were not larger than 100 μℓ. A 94.8% recovery was realized with a relative standard deviation of 2.2%.[28]

4. Post-Column Derivatization

An alternative to derivatization before separation involves modifying terpenes after chromatography, thereby allowing application of already-known separation systems. A fluorigenic procedure has been developed for cardenolide triterpenes for detection after HPLC.[29] It is based on the interaction between hydrochloric acid and the steroid moiety of the cardiac glycosides that most likely leads to dehydration. A hydrogen peroxide-ascorbic acid mixture was added for enhancement of fluorescence. Fluorescence was produced with light at 350 nm, and emission occurred at 485 nm. The slow kinetics of this reaction (10 min) require longer reaction times and reaction units (longer spirals) with accompanying difficulties such as loss of resolution and large pressure drops. To overcome these drawbacks, the air segmentation principle used in connection with an autoanalyzer were adopted. In this manner a 100-fold increase in sensitivity over UV detection was achieved. The reproducibility of the derivatization step was 1.2% (relative standard deviation) and peak broadening was found to be 15%.

5. Direct and Reverse-Phase HPLC

The complexity of the triterpene constituents in extracts from higher plants has dictated the development of techniques with greater resolving potential. A recent HPLC method[30] entailed initial separation on a column by gel permeation followed by reverse-phase chromatography. A loop injection device on the second column permitted reconcentration of fractions of several milliliters and injection without loss of resolution. This part of the process was automatically controlled and solvent gradients could also be used on the second column if desired.

6. Ecdysones

The ecdysones are a family of moulting hormones found in plants (phytoecdysones) and insects (zooecdysones). They belong to the sterol group and as such have been readily separated by techniques such as GC, TLC, and HPLC (Figure 7). Prepurification of these compounds is normally required before successful resolution can be accomplished. Zooecdysones are particularly troublesome because of low yields and co-occurrence of many other lipids.

a. Phytoecdysones Extracts

Methyl ethyl ketone was used to extract root material. The solvent was then removed by vacuum distillation at the end of which isobutyl acetate was added and distillation continued to ensure complete removal of the original solvent. Carplex No. 80 (silicic acid) and anhydrous sodium sulfate were then added to the residual isobutyl acetate solution and the phytoecdysones were absorbed onto the Carplex. After removal of the solvent by filtration and washing with isobutyl acetate the solid residue was packed into a column and eluted with methyl ethyl ketone. The crude phytoecdysones eluted in this manner were subjected to alumina chromatography and recrystallized from ethyl acetate-methanol (10:1 v/v).[32]

b. Zooecdysone Extraction

To ten pupae in a beaker were added 20 mℓ of acetone containing dry ice. The mixture was dried under reduced pressure and at >50°C. Cyasterone (5 μg) and sea sand (15 g) were combined with the pupae and ground in a mortar to a fine powder. After Soxhlet extraction with 100 mℓ tetrahydrofuran for 24 hr the tetrahydrofuran extract was reduced to $^1/_2$ volume, and 3 g of Carplex No. 80 (silicic acid) was added. The solvent was removed on a rotary evaporator and the impregnated silicic acid was then subjected to Soxhlet extraction successively for 1 hr with 50 mℓ of each of the following solvents: *n*-hexane, benzene, ether, and tetrahydrofuran. The tetrahydrofuran extract was chromatographed over 3 g of silica gel (Merck silica gel 60) in a glass filter (15 AG-3). The column was washed with tetrahydrofuran and the eluents were reduced *in vacuo*. The residues was fractionated on a silica gel plate (20 × 20 cm, 500 μm thickness) developed with chloroform-ethanol

1 = 2β,14α-Dihydroxy-5β-cholest-7-en-3,6-dione

2 = 2β,3β,14α-Trihydroxy-5α-cholest-7-en-6-one

3 = 2β,3β,14α-Trihydroxy-5β-cholest-7-en-6-one

4 = 22-Deoxyecdysone = 2β,3β,14α,25-Tetrahydroxy-5β-cholest-7-en-6-one

5 = 5α-α-Ecdysone = 2β,3β,14α,22R,25-Pentahydroxy-5α-cholest-7-en-6-one

6 = α-Ecdysone = 2β,3β,14α,22R,25-Pentahydroxy-5β-cholest-7-en-6-one

7 = 5α-20-Hydroxyecdysone = 2β,3β,14α,20R,22R,25-Hexahydroxy-5α-cholest-7-en-6-one

8 = 20-Hydroxyecdysone = 2β,3β,14α,20R,22R,25-Hexahydroxy-5β-cholest-7-en-6-one

9 = 26-Hydroxyecdysone = 2β,3β,14α,22R,25,26-Hexahydroxy-5β-cholest-7-en-6-one

	R_1	R_2	R_3
5	α-H	H	H
6	β-H	H	H
7	α-H	OH	H
8	β-H	OH	H
9	β-H	H	OH

	R_1	R_2
2	α-H	H
3	β-H	H
4	β-H	OH

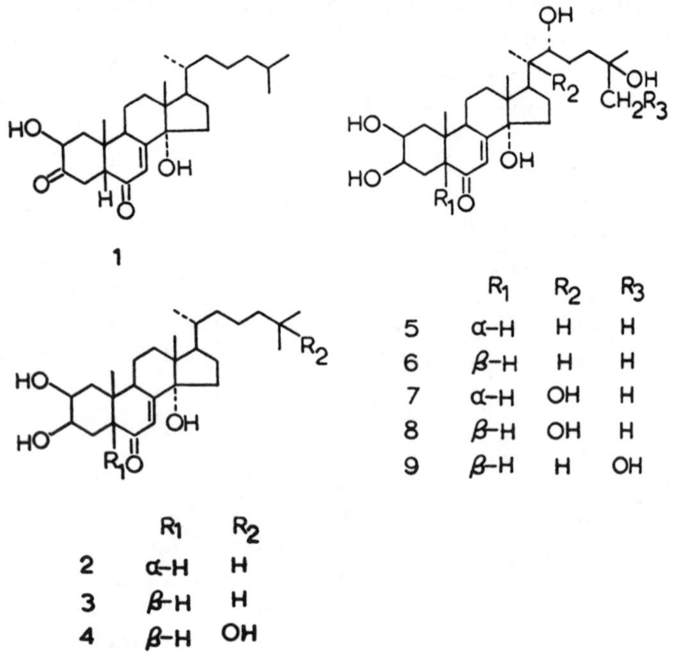

A

FIGURE 7. Ecdysone triterpenes. (A) Structures of the ecdysones resolved. (B) Elution profiles. The numbers over the peaks refer to compounds in 7A.

Instrument: Dupont 830 Liquid Chromatograph with UV detector 254 nm.

Column: 1 m × 2.0 mm I.D. SS Corasil II (37-50 μm).

Solvent: top trace - chloroform-ethanol (4:1); middle trace - chloroform-ethanol (9:1); and bottom trace - chloroform-ethanol (14:1)

Flow rate: 1.5 mℓ/min ambient temperature.[31]

(2:1). The ecdysones migrated in a broad band with an R_F range of 0.45 to 0.90. This was scraped off the plate and extracted with tetrahydrofuran using an ultrasonic generator, which afforded a recovery of 75%. The tetrahydrofuran was evaporated to dryness and the residue treated with 100 μℓ trimethylsilylimidazole. After heating the solution at 100°C for 30 min it was analyzed by mass fragmentography.[33]

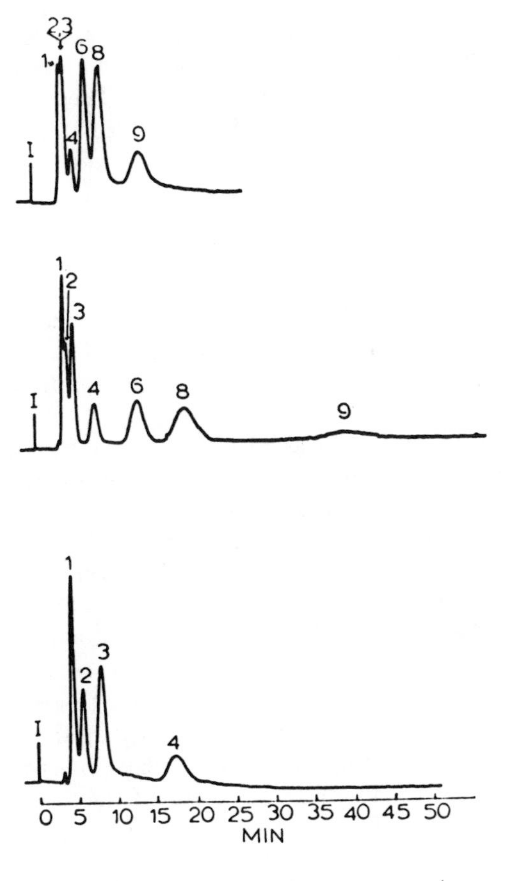

FIGURE 7B.

7. Acetates and Propionates

Triterpene acetates are prepared in the usual way. Thus, 1 to 20 μg of triterpene was dissolved in 0.5 mℓ anhydrous pyridine and after addition of 0.5 mℓ acetic anhydride, the mixture was kept in a ground glass-stoppered flask overnight at room temperature.

Propionates were prepared by treating the dry extract with 0.5 mℓ propionyl chloride and heating several minutes to complete dissolving. After 10 min the mixture was extracted with hexane and subsequently washed with water, 10% aqueous sodium bicarbonate, and water.[34] Acylation of epimeric alcohols may, in certain cases, facilitate their resolutions by TLC or GC.

E. POLYPRENOLS

1. General Aspects

Long chain polyisoprenoid alcohols are probably the most ubiquitous terpenes in nature. These monohydric alcohols vary in length depending on the organism. The simplest members of this class containing 1, 2, or 3 isoprenyl units linked in a head-to-tail fashion are most prevalent in plants. Polyprenols have been found in varying lengths in bacteria, plants, fungi, and mammals, the latter two containing the largest representatives. For example, dolichols in porcine liver contain 17 to 22 isoprene residues, while those in the fungus *Aspergillus fumigatus* have up to 24 isoprene units. Unlike carotenoids many of the polyprenols are partially saturated imparting a greater stability. Of the saturated derivatives examined thus far, all contain a saturated α-isoprene residue (OH-containing). Many polyprenol internal double bonds are *cis* rather than *trans*. The *trans* isomers are usually found in isoprenoid quinones. Thus a great diversity of size, stereoisomers, and degree of saturation is seen in this class, even in populations from a single cell type (Figure 8). In carrying out cellular functions, polyprenols occur in mono- and diphosphorylated forms, and in turn sugar residues are attached to the phosphate in glycan biosynthesis.[36]

2. Isolation

a. Polyprenols from Plants by Saponification

The isolation of polyprenols is similar to that of many nonpolar terpenes. Tissue is digested by alkaline saponification in the presence of pyrogallol and the nonsaponifiable fraction is then extracted usually with ethyl ether and the latter is fractionated on alumina, silica gel, Florisil, or Sephadex LH-20.[37]

For plants, 250 g of leaf tissue was macerated and refluxed for 1 hr in a mixture of 60% aqueous potassium hydroxide (750 mℓ) and 5% pyrogallol in methanol (1500 mℓ). After filtering and diluting with water, the aqueous solution was extracted with 3 × 1500 mℓ diethyl ether. The ether extract was washed free of base and dried. Polar impurities were removed by slurrying 500 mℓ of a 25% diethyl ether in light petroleum solution of the extract with Brockmann grade 3 alumina (375 g) in light petroleum (250 mℓ) and filtering after vigorous shaking. The alumina was then eluted with 2 × 500 mℓ of 25% ethyl ether in light petroleum. The filtrate was directly taken up in light petroleum and the sterol removed by two crystallizations from this solution at 0°C. The mother liquors were chromatographed on alumina, Brockmann grade 3, whereupon 3% ether in light petroleum eluted hydrocarbon and plastoquinones and 10% ether in light petroleum eluents contained prenols, tocopherols, and ubiquinones. Acetylation of this eluate in the usual fashion with acetic anhydride and pyridine was followed by elution from alumina, Brockmann grade 3, with a step gradient of light petroleum followed by 0.25%, 0.5%, and 1.0% ether in light petroleum. The acetates were then saponified back to the free alcohol using conditions comparable to those of the original saponification.[38]

b. Polyprenols from Mammalian Tissue

Procedures for isolation of polyprenols from liver, kidney, and brain follow essentially the same procedure used for plants.[39] For example, tissue was saponified by refluxing in 1 volume of a solution of 5 *N* KOH in 50% aqueous ethanol containing pyrogallol (10 mg/kg tissue) for 1 hr under a nitrogen atmosphere. The nonsaponifiable fraction was extracted with 3 × 0.3 volumes of ethyl ether and the organic phase was combined diluted with 0.3 volumes of petroleum ether and washed with 6 × 0.2 volumes of water. After evaporation of the ether *in vacuo* at 37°C, the residue was dissolved in a minimal amount of boiling hexane. Upon standing overnight at 4°C, the bulk of the sterols precipitated and the supernatant was collected by filtration and reduced in volume.

The hexane soluble fraction was chromatographed on a column of silicic acid (20 mg lipid per g silicic acid) eluting with 200 mℓ of 2% ether in petroleum ether and then 300 mℓ of 4% ethyl ether in petroleum ether. The column effluent was monitored by TLC. Mammalian polyprenols were eluted in the fractions obtained with the last 200 mℓ of 4% ethyl ether in petroleum ether.[40] These fractions were combined and evaporated to dryness. Pure polyisoprenols were isolated from this fraction by preparative thin-layer chromatography.[41]

Another procedure has been developed for the isolation of polyprenols from bovine thyroid glands. It consists of the following steps. Tissue, freed of adipose and connective tissue, was extracted with 3 × 10 volumes of acetone at 4°C under nitrogen with stirring. The efficiency of extraction was monitored by a fourth extraction with chloroform-methanol (2:1). The acetone extract was evaporated to dryness and then chromatographed on a silicic

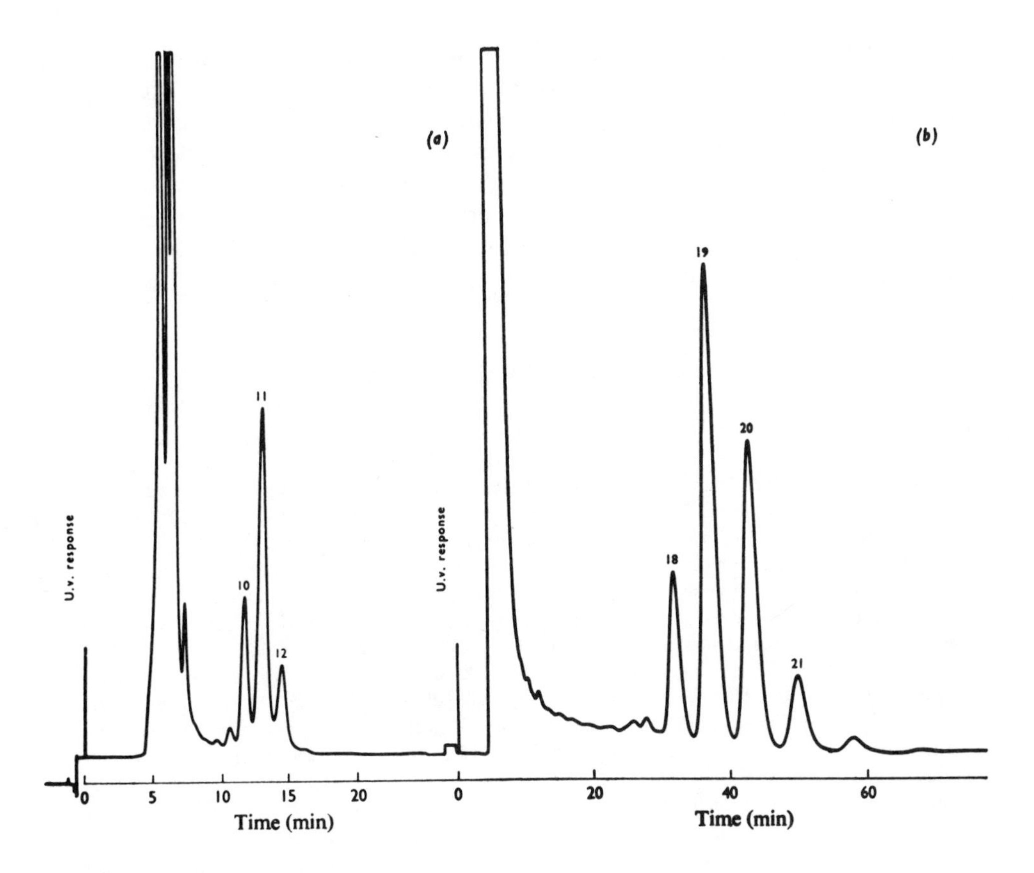

FIGURE 8. HPLC separation of polyprenols.
Instrument: Varian model 1400 HPLC equipped with a Glenco valve or a Waters model 6000A HPLC.
Column: μBondapak C_{18} reverse-phase column (30 cm × 4 mm I.D.) (Waters Assoc.).
Solvent, flow rate: propan-1-ol-water (95:5), 0.5 mℓ/min.
Compounds: The numbers on the major peaks refer to the total number of isoprene units.[35]

acid column (a 2.5 × 100 cm column was used for 2 kg of tissue). After eluting with one column volume of hexane and four volumes of linear gradient of 0 to 25% diethyl ether in *n*-hexane, the fractions containing the dolichols and dolichol esters were detected by TLC [developing solvent: light petroleum (b.p. 60 to 80°C)diethyl ether-acetic acid, 70:30:1 v/v; detection:iodine]. The same TLC conditions were used for preparative TLC of dolichol after combining appropriate fractions. The purification of dolichol esters was accomplished by preparative TLC using light petroleum-diethyl ether-acetic acid (90:10:1). The esters were best detected by staining a small section of the plate with Rhodamine-6G spray. The dolichol and dolichol ester bands were scraped from the TLC plates and eluted with chloroform.[42]

c. Bactoprenols from Lactobacillus casei

Harvested washed cells from a 24-ℓ batch were added to 500 mℓ of 80% (v/v) ethanol and the suspension was distilled until the volume was reduced to about 100 mℓ. After adding an equal volume of water, the mixture was extracted with 3 × 200 mℓ of diethyl ether. Phase separation was facilitated by centrifugation. The ether extract was dried over sodium sulfate and then evaporated *in vacuo*. The residue was taken up in 5 mℓ of hot ethanol and permitted to stand ovenight. The ethanol solution was then filtered to remove precipitated lipid and evaporated to dryness *in vacuo*.

The residue was taken up in a minimum amount of *n*-heptane and added to a column of alumina that was deactivated with acetic acid.[43] Polyprenols were eluted with 280 mℓ of 20% (v/v) benzene in *n*-heptane. This fraction was evaporated to dryness and the residue acetylated in 0.4 mℓ of pyridine and 0.2 mℓ of acetic anhydride at room temperature overnight. The reagents were then removed with a stream of nitrogen and the acetylated product chromatographed on a second column of acetic acid-deactivated alumina. This column was eluted with 140 mℓ of *n*-heptane that had been freed from ketones by distillation over 2,4-dinitrophenylhydrazine (4 g) and 1 mℓ of conc. HCl per liter. The polyprenols were found in the last 100 mℓ of eluent.[44]

*d. Polyprenol Phosphates and Pyrophosphates from Bacteria**

The work-up of the cells of *Micrococcus lysodeikticus* was carried out in three equal parts as follows. Cells (220 g) were mixed with 100 mℓ of water and a slurry was prepared. Acetone (2.5 ℓ) at −20° was added. The mixture was stirred for 10 min and then filtered through a sintered glass funnel. The residue was extracted by mixing with 750 mℓ of chloroform-methanol (2:1) at room temperature for 20 min. The material was filtered and the residue was again extracted, this time with 500 mℓ of butanol-1-pyridinium acetate, pH 4.2 (2:1). The extraction with the latter solvent was facilitated by subjecting the suspension to sonification in a Branson sonifier in the presence of 100 mℓ (packed volume) of washed glass beads (5 μ diameter, Heat Systems Ultrasonics, Plainfield, NY). The extraction with butanol-pyridinium acetate was repeated two more times, and the butanol-pyridinium acetate extracts were combined. This extract was washed four times each with 0.5 volumes of water, and the resulting aqueous phases were then backwashed once with butanol-1. The butanol layers were combined and approximately 150 mℓ of pyridine were added to raise the pH before concentration to dryness on a rotary evaporator. The solids obtained from the three separate runs on this scale were combined. Yields of solids in the three fractions were then Fraction 1, acetone extract, 1.1 g; Fraction 2, chloroform-methanol extract, 5.1 g; and Fraction 3, butanol-pyridinium acetate extract, 3.2 g.

Separately, on a 1-ℓ scale, cells (*Micrococcus lysodeikticus*) were grown to half-maximum and then transferred to medium containing 1% Difco Bactopeptone, 0.01% yeast extract, 0.5% NaCl adjusted to pH 7.5 to which inorganic phosphate (25 mCi) was added last. After 3 hr, bacitracin (160 μg per mℓ) was added, and the cells were harvested 3 hr later. The lipids were prepared from the small scale experiment in the same manner as described above. Each fraction was dissolved in 1 mℓ of carbon tetrachloride to which were added 9.4 mℓ of ethanol, 0.8 mℓ of water, and 0.3 mℓ of 1 *N* NaOH (final concentration, 0.026 *N*). This mixture was incubated at 37°C for 30 min after which time the pH was still alkaline. Ethyl formate, 0.5 mℓ, was added, and the mixture was incubated for 5 min at 37°C. After this treatment, the pH was approximately 7. The sample was taken to dryness on a rotary evaporator, dissolved in 6 mℓ of chloroform-methanol, 2:1, and 4 mℓ of water were added. The lower organic phase was washed twice with 5 mℓ of methanol-water, 1:1, and the aqueous washings were backwashed once with chloroform. The organic phase was finally taken to dryness on a rotary evaporator. This procedure converts phosphatides to water-soluble products, leaving the prenylphosphates as the main components of the organic phase. The total counts before alkaline hydrolysis in each of the three fractions were as follows: Fraction 1, acetone extract, 2.1×10^8 cpm; Fraction 2, chloroform-methanol extract, 4.4×10^8 cpm; and Fraction 3, butanol-pyridinium acetate extract, 6.5×10^8 cpm. The total counts after alkaline hydrolysis in each of the fractions was as follows: Fraction 1, 3.1×10^5 cpm (0.15%); Fraction 2, 28.5% $\times 10^5$ cpm (0.64%); Fraction 3, 564×10^5 cpm (8.7%). It is clear, therefore, that whereas substantial amounts of phospholipids are extracted by each solvent, about 95% of the prenylphosphates are extracted by butanol-pyridinium acetate. Thin layer chromatography, followed by radioautography, indicated that the major radioactive component of each of the saponified fractions was a compound with the mobility of lipid pyrophosphate. Since Fractions 1 and 2 were relatively small compared to that found in Fraction 3, only Fraction 3 from the radioactive experiment and from the large scale isolation was worked up further.

The total material from Fraction 3 of the large scale isolation (3.2 g) was dissolved in 20 mℓ of carbon tetrachloride to which were added 180 mℓ of ethanol, 16 mℓ of water, and 6 mℓ of 1 *N* NaOH. The mixture was incubated at 37° for 30 min, following which 10 mℓ of ethyl formate were added, and the mixture was again incubated at 37° for 5 min. The mixture was then taken to dryness on a rotary evaporator, dissolved in 800 mℓ of chloroform-methanol, 2:1, to which were added 400 mℓ of water. The organic phase was washed three times with 50% methanol and the aqueous washings were backwashed once with chloroform. The combined chloroform phases were taken to dryness (yield, 1.24 g of a yellow-brown oil).

* **Stone, K. J. and Strominger, J. L.,** *J. Biol. Chem.*, 247, 5107, 1972. With permission.

This material was then mixed with the corresponding fraction obtained from the radioactive experiment. The pooled material suspended in 20 mℓ of acetone was applied to a column of Silica Gel G (Woelm, 40 × 3.7 cm, 250 g). The silica gel had previously been extensively washed in acetone and fines discarded. The column was packed using acetone. After application of the sample, the column was washed with 1 liter of acetone which was collected in one fraction. The column was then eluted at a rate of about 1.5 mℓ per min, collecting 14 mℓ per tube, with the following: Solvent 1, chloroform-methanol, 2:1, 1 liter; Solvent 2, absolute methanol, 1 liter; Solvent 3, methanol containing 2% water, 1 liter; and Solvent 4, butanol-6 *M* pyridinium acetate pH 4.2, 2:1, 1 liter. The radioactive material was eluted in two peaks. However, previous experiments indicated that C_{55}-prenylpyrophosphate readily forms complexes with less polar lipids, and the first peak was believed to represent such a complex. The yields in the two peaks were in Peak 1, 1.9×10^7 cpm, and in Peak 2, 3.5×10^7 (corrected for radioactive decay), together representing over 95% of the radioactivity applied to the column.

The two peaks were pooled (240 mg, 5.4×10^7 cpm). The material was then applied to a column of Silica Gel H (Woelm 30 g, packed and washed in chloroform). The column was eluted with a linear gradient obtained with 565 mℓ of chloroform-methanol (18:1) in the mixing vessel, and 725 mℓ of chloroform-methanol-water (65:25:4) in the reservoir. The sample was applied in a small volume of the same mixture as was in the mixing vessel. The elution profile indicated the presence of a single major radioactive component. Other nonradioactive lipids were detected by their yellow color (absorbance at 425 nm).

The material in the peak tubes was pooled and taken to dryness on a rotary evaporator (29.9 mg, 3.3×10^7 cpm). Approximately 25% additional material was obtained by reprocessing the material from the leading and trailing edges of the main peak by the same procedure. This material (0.86×10^7 cpm) was then pooled with the main fraction (total, 4.2×10^7 cpm). This pooled material was then applied to another column of Silica gel H (8 g) and a shallower gradient was employed for development. Chloroform-methanol (95:5), 166 mℓ, was in the mixing chamber and chloroform-methanol-water (65:35:4), 200 mℓ, the reservoir. Fractions of 2.7 mℓ were collected, and a broad peak of radioactivity was obtained which appeared to be separated from the material absorbing at 425 nm. The main peak was pooled and taken to dryness on a rotary evaporator (24.0 mg, 3.44×10^7 cpm, 82% yield).

This material was a yellow oil. In order to purify it further, it was dissolved in 0.4 mℓ of chloroform-methanol (2:1), and 5 mℓ of absolute ethanol were added. The material became cloudy and was put in an ice bath for 5 min. A white precipitate was removed by centrifugation (12.1 mg, 3.36×10^7 cpm, 60% yield from the starting material). The yellow supernatant solution contained very little radioactivity (12.4 mg, 0.06×10^7 cpm) and was discarded. Repeating the precipitation as above again yielded a white precipitate (12.1 mg) without loss of radioactivity.

*e. Polyprenol Phosphate Oligosaccharides**

Total lipid was isolated from *Micrococcus lysodeikticus* cells by extraction in 10 volumes of chloroform-methanol (2:1). The mixture was stirred at room temperature for 4 to 12 hr after which time insoluble material was removed by filtration. The filtrate was washed with 0.2 vol of 0.9% aqueous sodium chloride. The chloroform extract was then concentrated.

The lipid (2.0×10^6 cpm, 8.15 μmoles of mannose) was applied to a silicic acid column (60 gm) and eluted successively with 1000 mℓ of $CHCl_3$, 600 mℓ of acetone, and 500 mℓ $CHCl_3$-CH_3OH, 1:1. Crude mannosyl-1-phosphorylpolyisoprenol (MMP), quantiatively recovered in the last fraction, was dissolved in $CHCl_3$-CH_3OH, 2:1 (CM) and applied to a DEAE-cellulose column (4.5 × 30 cm) prepared in CM. The column was eluted with 2400 mℓ of CM, 700 mℓ of CH_3OH, and 700 mℓ of CM containing 84 mℓ of conc. NH_4OH. All of the radioactive lipid was recovered in the acidic lipid fraction eluted with the last solvent. This fraction was evaporated to dryness and dissolved in 16 mℓ of $CHCl_3$-CH_3OH, 1:4. After addition of 1.5 mℓ of 1 *N* NaOH, the solution was incubated at 37° for 15 min. Then 1.5 mℓ of 1 *N* acetic acid, 30 mℓ of $CHCl_3$-CH_3OH, 9:1, 15 mℓ of isobutanol, and 30 mℓ of H_2O were added and the solution was mixed vigorously. The aqueous layer was discarded and the $CHCl_3$ layer was washed with 15 mℓ of H_2O-CH_3OH, 2:1, and then evaporated to dryness. The C^{14}-MPP (1.72×10^6 cpm, 7.0 μmoles) was applied to a DEAE-cellulose column and eluted with CM AND CH_3OH as described above. The column was then eluted with 0.005 *M* ammonium acetate in 99% CH_3OH and 10-fractions were collected. MPP was eluted in a single radioactive peak between fractions 144-160. The pooled fractions were evaporated to dryness, dissolved in 2 mℓ of 1% CH_3OH in $CHCl_3$, and applied to a Unisil silicic acid column (1 × 7 cm). The column was eluted with 10 mℓ each of 1% CH_3OH in $CHCl_3$, 5% CH_3OH in $CHCl_3$, and 50% CH_3OH in $CHCl_3$. The last effluent contained 1.28×10^6 cpm (5.2 μmoles) of MPP. Examination of MPP by thin-layer chromatography on silica gel (eluent: $CHCl_3$-CH_3OH-H_2O, 12:6:1) revealed the presence of one compound that was positive in tests for PO_4 and lipid (Rhodamine). This compound (R_f = 0.22) contained greater than 95% of the radioactivity. Two other minor components (R_F = 0.48, 0.58) were detected with Rhodamine; both were free of PO_4 and radioactivity.

* **Scher, M., Lennarz, W. J., and Sweeley, C. C.,** *Proc. Natl. Acad. Sci. U.S.A.*, 59, 1313, 1968.

MPP was further purified by gel filtration on a Sephadex column. To a column (2.5 × 81 cm) of LH-20 Sephadex prepared in 0.01 *M* ammonium acetate in $CHCl_3$-CH_3OH, 1:1, was added 0.643 × 10^6 cpm (2.62 μmoles) of MPP. The column was eluted with the above solvent at a flow rate of 0.5 mℓ/min and 2-mℓ fractions were collected. The MPP was eluted in a single radioactive peak in fractions 100-108 and 0.575 × 10^6 cpm (2.43 μmoles of MPP) was recovered.

3. Derivatization of Polyprenols

In addition to acetylation (see above), the single hydroxyl of the polyprenols has been derivatized with trimethylsilyl, trifluoroacetyl, and nitrobenzoyl chlorides following routine procedures.

4. Hydrogenation

Polyprenols and their acetates can be hydrogenated to facilitate their separation by GC.[47,48] The derivatives were readily hydrogenated at room temperature and pressure in a Towers microhydrogenation apparatus using platinum oxide as catalyst and cylcohexane as solvent for acetates, whereas for the free alcohols, cyclohexane-ethanol-acetate acid (1:1:1) was used.

5. Ozonolysis

The castaprenol mixture (42 mg) was acetylated and dissolved in 5 mℓ of ethyl acetate that had been redistilled over 2,4-dinitrophenylhydrazine. Ozonization was carried out for 30 min at 0°C. Thereupon the solution was gassed with oxygen until ozone could no longer be detected with starch-iodide paper. The ozonides were reduced by treatment with zinc dust (50 mg) and acetic acid (1 drop) for 16 hr. The suspension was filtered into a solution of 2,4-dinitrophenylhydrazine (1 g freshly extracted with light petroleum) in 2 *N* HCl and the 2,4-dinitrophenylhydrazones sedimented upon standing. A blank using ethyl acetate alone afforded no precipitates, indicating that the hydrazine derivatives were from castaprenols.

After extraction with light petroleum in a Soxhlet appartus for 3 hr, the ultraviolet absorption spectra of the soluble and insoluble fractions were compared with those of authentic compounds.[38,39]

F. CAROTENOIDS

1. General Aspects

Although most carotenoids are molecules containing 40-carbon atoms, C_{20}-, C_{25}-, C_{45}-, and C_{50}-compounds have been isolated from certain plants and bacteria. The acyclic highly unsaturated nature of carotenoids makes them nonvolatile and/or thermally unstable as well as sensitive to light, acid and oxygen. Unless they are hydrogenated, they are not suitable for GC. While early techniques of liquid-column chromatography e.g., alumina (Figure 9) MgO-Supercel, $ZnCO_3$, $CaCO_3$, $Ca(OH)_2$ for less polar carotenoids, and cellulose and sucrose for xanthophylls, are adequate for large scale preparative work and for resolving carotenoid hydrocarbons from their oxygenated metabolites, they do not permit the chromatography of molecules differing only by the number of isoprene units. For this purpose TLC has superseded paper chromatography, and it is still one of the most prevalent methods presently used (see Table TLC 22-29). Recently HPLC has proven far more suitable, especially for the resolution of isoprenologs in bacteria and plants.[50]

Indeed since HPLC of carotenoids can be readily performed without undue exposure to light or oxygen, it is rapidly becoming the method of preference. The technique is also superior to column chromatography or TLC in speed and in most cases in resolving ability, particularly for isomeric mixtures. One of the early applications of HPLC was to resolve citrus carotenoids.[51]

Regardless of the technique used, certain general precautions must be taken to avoid a loss of labile carotenoids and the production of artefacts. Procedures which entail excessive

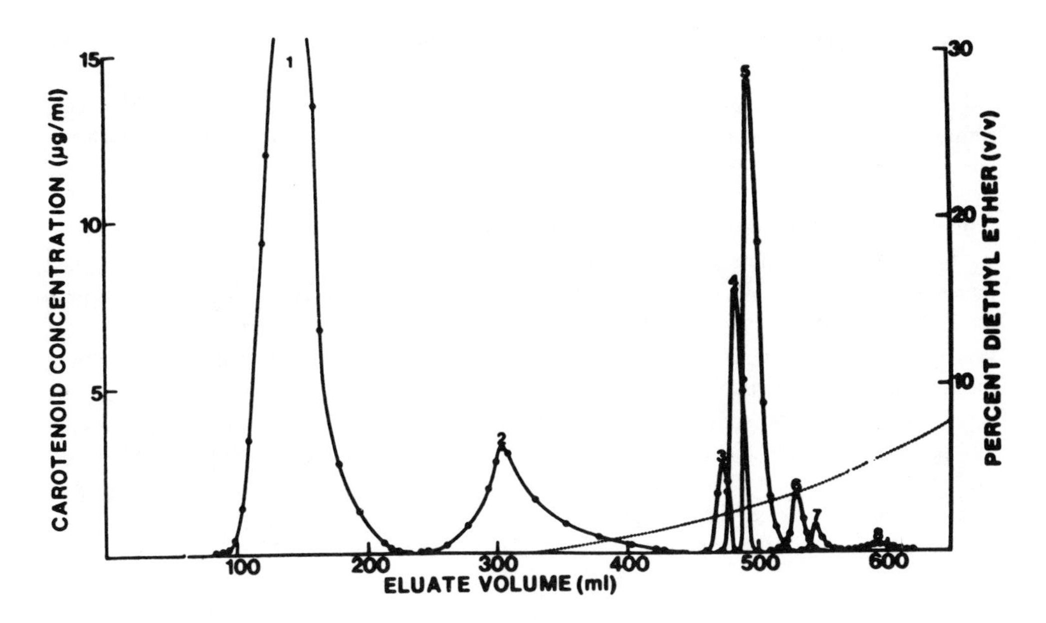

FIGURE 9. Separation of neutral carotenoids of *Neurospora crassa* using concave gradient elution from deactivated alumina.

Column: 30 × 2 cm alumina (80-120 mesh, 6% water).

Solvent: Light petroleum followed by an increasing concave gradient of diethyl ether in light petroleum (dotted line).

Compounds: Compounds were monitored using a flow through cell in a Beckman Acta III spectrophotometer set at 450 nm. Fractions containing two carotenoids were resolved using Beer's Law for two component mixtures. 1=phytoene, 2=phytofluene, 3=ζ-carotene, 4=β-carotene, 5=neurosporene, 6=torulene, 7=lycopene, 8=3,4-dehydrolycopene.[49]

heating, or the presence of air or light, during chromatography lead to *cis-trans*-isomerization, epoxidation, and general degradation. Therefore, chromatography tanks and columns should be wrapped in aluminum foil and general manipulations carried out in subdued light. It is not recommended that evaporation *in vacuo* be performed at temperatures above 40°C. Carotenoids are readily oxidized in air; and particularly susceptible compounds should be subjected to TLC in a nitrogen atmosphere. When evaporating solutions of carotenoids to dryness it is recommended that 99.9% purity nitrogen be used, and for some carotenoids the gas should be oxygen-free and dry. Like some lower terpenes, carotenoids are acid-labile, and conditions of low pH should be avoided during isolation. Solvents should be peroxide-free, anhydrous, and redistilled. Methylene chloride is preferred over chloroform because the latter often contains traces of HCl in it. An excellent review on specific procedures for carotenoid isolation and purification has been written by B. H. Davies.[52] Chemical modification of the carotenoids are described in the same chapter and preceeding sections[53] as well as in earlier books.[54] The structural diversity of carotenoids is such that there are now over 400 different naturally occurring representatives of this group.

2. Sample Preparation

a. Simple Extraction of Plants

Fresh spinach was stripped of main stems and veins and blended in 6 volumes of acetone [oxidation by air can be reduced by the use of N_2 or CO_2 (dry ice) atmosphere]. The acetone extract was reduced to 10 mℓ and redissolved in diethyl ether (100 mℓ). The ether solution was extracted with 6 × 50 mℓ of 10% aqueous NaCl to remove some lipid and acetone. The ether extract was dried over sodium sulfate and evaporated to dryness. The sample was taken up in methanol (6 mℓ) with several drops of acetone to completely dissolve the residue for chromatography.[55]

For carotenoids from tomato, methanol-isooctane-cyclohexane extractions have been performed.[56] Algal carotenoids have been purified by solvent partitioning between acetone-water and petroleum ether.[57]

b. Extraction, Solvent Partitioning, and Saponification of Halobacterial Lipids

Total lipids were subjected to acetone precipitation and then supernatant that contained pigments, squalenes, and vitamin K was evaporated to dryness on a rotary evaporator in the dark. The deep red residue (180 mg from 50 ℓ culture) was kept under nitrogen until it was treated with 95% methanol and hexane (50 mℓ each) in a separatory flask. After removing the hexane layer the methanol was washed with 3 × 50 mℓ of hexane and the hexane extract combined. The hexane solution was then washed with water to remove methanol and then evaporated to dryness under nitrogen. The residue was saponified by treatment with 10% methanolic KOH for 4 hr at room temperature. The mixture was extracted with hexane, back-washed with water until neutral to phenolphthalein, dried over anhydrous sodium sulfate, and reduced to 5 mℓ with a stream of oxygen-free nitrogen.[58]

c. Isolation from Fungi and Yeasts

Although carotenoids can be extracted from most fungi and yeasts following the same procedures developed for higher plants, there are some yeasts that are difficult to extract because of their tough cell walls. Initial mechanical disruption of these cells using a French press or glass beads in a colloid mill or homogenizer is then necessary.[52]

d. Determination of the Carotenoid, Phytoene, in Blood

A method based on HPLC has been developed to detect orally administered phytoene in blood. An overall recovery of 86 ± 6% was realized and the limit of detection was 50 to 100 ng/mℓ blood. The method serves primarily to detect phytoene in the presence of dietary β-carotene, which is a precursor of vitamin A.

Whole blood (1 mℓ) and saline (2.0 mℓ) were added to a 50 mℓ centrifuge tube (PTFE No. 16, stoppered) and mixed thoroughly on a Vortex mixer. Five control blood specimens containing 0, 250, 500, 750, and 1000 mg of phytoene were analyzed with the unknowns. After addition of 2.5 mℓ of absolute ethanol and occasional agitation, protein precipitation occurred (about 5 min). A 10 mℓ solution of *n*-hexane-isopropanol (95:5) was added and the tube placed on a reciprocating shaker (Eberbach, Ann Arbor, Mich.) at 80 to 100 strokes/min for 10 min. The samples were spun at 2500 rpm (1500 *g*) in a refrigerated centrifuge (Model PR-J, rotor No. 253, Damon/IEC, Needham, Mass.) at 5°C. A 9 mℓ aliquot of the organic phase was transfered into a 15 mℓ conical centrifuge tube and washed with 1 mℓ of water (5 min on the reciprocal shaker) followed by centrifugation at 2500 rpm as above for 5 min. The lower aqueous layer was removed with a hypodermic syringe fitted with a 20 gauge 6 in cannula (Becton Dickinson, Rutherford, N.J.). The 1 mℓ water wash was repeated if a heavy lipid layer remained at the interphase. An 8 mℓ aliquot of the organic phase was transferred into a 15 mℓ conical centrifuge tube and evaporated to dryness at 60°C in a N-EVAP evaporator (Organomation Associates, Worcester, Mass.) with a stream of dry nitrogen. The residue was taken up in 100 μℓ of isopropanol and 5 to 10 μℓ aliquot were used for HPLC analysis.

Analyses were performed with a 0.5 m × 2 mm I.D. stainless-steel column containing 1% ODS Permaphase chemically bonded on Zipax (DuPont, Wilmington, Del.) in a DuPont Model 830 high-pressure liquid chromatograph equipped with a Model 836 multi-wavelength UV and fluorescence detector (operated in the UV mode at 280 nm). The isocratic mobile phase was a mixture of water-methanol (5:95) pumped at a flow-rate of 0.8 mℓ/min at ambient temperature. Under these conditions the retention time of phytoene was 2.5 min and that of β-carotene was 4.9 min.[59]

e. Removal of Sterols

Sterols can be a major component of the unsaponifiable fraction containing carotenoids and can be precipitated out of a light petroleum solution by allowing it to stand overnight at −20°C. Another approach involves precipitating sterols out as their digitonides.

Unsaponifiable material (20 mg) was dissolved in 6 mℓ of aq. 95% ethanol in a centrifuge tube and heated to boiling; to this was added 5 mℓ of a boiling solution containing 100 mg digitonin in aq. 90% ethanol. The mixture was boiled until a white precipitate of digitonides appeared. Precipitation was completed overnight at 4°C and the mixture was centrifuged and the supernatant retained. The digitonides were washed free from pigment with precooled ether, which was then combined with the ethanolic supernatant in a small separatory funnel. The carotenoids were transferred to the ether phase by the addition of water and the sterol-free unsaponifiable fraction was recovered from the ether by washing with water, drying and evaporation of solvent.[52]

3. Hydrogenation of Carotenoids

Low pressure hydrogenation of carotenoids affords more volatile derivatives that are resolvable by GC. Phytoene, lycopene and phytofluene were reduced in a Parr hydrogenation apparatus with platinum oxide as catalyst at a pressure of 40 psi with petroleum ether-isopropyl alcohol as solvent.[60]

For small samples hydrogenations were performed in a Gallenkamp microhydrogenator apparatus (A. Gallenkamp, London, England). Carotenoids (50 to 500 μg) were dissolved in 25 mℓ of chloroform containing 50 mg of platinum oxide and were shaken at room temperature for 2 to 4 hr under a positive hydrogen pressure of 100 mmHg. Catalyst was removed by filtration through a sintered glass funnel and the filtrate evaporated *in vacuo* to a final concentration of approximately 10 μg/μℓ chloroform. A 1 to 3 μℓ aliquot was injected onto a GC column.[61]

4. Modification of Hydroxyl Groups

a. O-Methylation of Carotenoids

Selective O-methylation of allylic secondary hydroxyl groups of carotenoids has been accomplished using hydrogen chloride in methanol whereas nonallylic hydroxyl groups are alkylated by methyl iodide and potassium t-amyl oxide or the Kuhn procedure (methyl iodide and silver oxide or barium oxide in dimethyl formamide). The latter procedure also affords methylated tertiary alcohols, but gives abnormal products as well. In the carotenoid series as with many other compounds, enolic and phenolic hydroxyl groups are not readily methylated with diazomethane.

A new method utilizing methyl iodide and sodium hydride was shown to be superior to the above methods for methylation of nonallylic and allylic secondary hydroxyl groups including sterically hindered ones. Enolic hydroxyl groups also undergo methylation, but iodinated side products are produced as well.

The protocol for this improved method involves dissolving 0.1 to 1 mg of carotenoid in dry tetrahydrofuran and adding freshly distilled methyl iodide (1 mℓ) and 40 mℓ sodium hydride. The reaction mixture is kept in the dark in an inert atmosphere at room temperature for 16 to 24 hr and worked up in the usual manner.[62]

b. Acylation and Silylation

Like other terpenoid primary and secondary hydroxyl groups carotenoid hydroxyls are readily acylated and silylated following general procedures. Tertiary hydroxyl functions are also derivatized by the latter method, but not by the acetic anhydride-pyridine procedure.

c. Oxidation

Allylic secondary alcohols are oxidized to ketones by *p*-chloranil, a reaction that causes a bathochromic shift in the visible light range.

The carotenoid (1 to 10 mg) was dissolved in benzene (5 to 50 mℓ) and ethanol (0.5 to 2 mℓ) and 3 to 30 mg of chloranil added. After bubbling nitrogen gas through the mixture for 10 min, iodine (1 to 5% of the carotenoid weight) in light petroleum (0.5 to 5 mℓ) was added. The solution was illuminated wth a filtered 200 W incandescent lamp at a distance of 70 cm.[63]

Other selective oxidizing agents including MnO_2, NiO_2, dichlorodicyanobenzoquinone, silver oxide, and iodine in light have been used.[54]

d. Dehydration

Mild acid treatment of carotenoids leads to elimination of water from allylic hydroxyl groups of carotenoids. Allylic ether groups such as glucosides are also displaced. Substitution patterns, in which a hydrogen is not available, preclude elimination using this method. The usual procedure entails treating the carotenoid in chloroform with a few drops of a saurated solution of HCl (0.3 *M*) in chloroform, the latter prepared by bubbling hydrogen chloride gas into the solvent for 3 to 5 min or by saturating chloroform with conc. HCl. The reaction takes about 15 min and is monitored spectrophotometrically. The additional double bond leads to a spectral shift of about 16 nm higher wavelength. Less quantitative dehydrations lead to shifts of 10 nm or more. Products are recovered by washing two times with $NaHCO_3$ solution and water.[54] Phosphorus oxychloride in pyridine has been used to dehydrate non-allylic tertiary alcohols.[54]

5. Carotenoid Glycosides

As in the case of monoterpene glycosides, these derivatives are polar and amphipathic. Acetylation or trimethylsilylation affords a more uniformly hydrophobic molecule. As mentioned above, acidification leads to deglycosylation. Carotenoid glycosides are stable to base, but unlike the monoterpenoid glycosides they are not readily hydrolyzed by commercially available α- and β-glucosidases; this is probably due to a lack of specificity for the large aglycone moiety.

6. Epoxides

Carotenoid epoxides are susceptible to acid and hydrogenolytic cleavage. If a 5,6-monoepoxide is dissolved in ethanol and a drop of conc. HCl added to this solution, a hypsochromic shift of 17 to 22 nm occurs. A shift of 40 nm is seen with a 5,6,5′,6′-diepoxide. If an etheral solution of carotenoids containing 5,6 or 5,8-epoxide groups is shaken with aqueous 20% HCl, a blue color is formed. Diepoxides give more stable colors and these are useful for TLC visualization and identification (see p. 114).

G. A VITAMERS

1. General Aspects

A vitamers are derived from β-carotene by a symmetrical cleavage of the C_{40} molecule in mammals.[64] Thus they contain a trimethylcyclohexene ring, a polyene side chain and a polar end group, the latter being either an alcohol, aldehyde or acid. In addition, this group is often found in a derivatized form, with esterification of the alcohol group being most common. Because of these structural features, vitamin A compounds behave much like carotenoids chromatographically. They are also labile, and all the precautions in handling enumerated for carotenoids (see Section III.F.1) apply to the A vitamers. For most A vitamers light and oxygen sensitivity are especially acute. In fact, the retinyl derivatives offer more of a challenge to the analytical chemist than carotenoids because they are present in trace concentrations in mammals under normal circumstances. Because of their important role in general growth, the growth and differentiation of epithelial tissue, and visual function and reproduction, a great deal of effort has been invested in detecting A vitamers in various body fluids and tissues in mammals.

2. Detection in Plasma

a. HPLC

Until recently, fluorescent and colorimetric methods of detection (see below) were the best available. These techniques suffer, however, from a lack of specificity and interference by carotenoids and other variable dietary constituents. In the last few years HPLC techniques have been developed that not only permit determination of serum and plasma retinol, the major vitamin A derivative in this fluid, but simultaneously, esters as well as other A vitamers. Initially, prefractionation on silica gel columns was performed.[65,66]

More recently the column of choice for HPLC has been octadecylsilica gel, thereby facilitating simultaneous determinations. On-line UV detection makes possible the measurement of 50 μg/ℓ of retinol and 100 μg/ℓ of retinyl esters in 200 μℓ of serum. Within-run precision was 2.3% for retinol and 4.3% for retinyl palmitate, while day-to-day precision was 4.9% during a month using the Bligh-Dyer extraction procedure as follows:

To a centrifuge tube was added 200 μℓ of serum or plasma, 0.6 mℓ of water, 2.0 mℓ of methanolic solution of internal standard (137 μg/ℓ) and 1 mℓ of chloroform. After mixing thoroughly for 1 min and allowing to stand 5 min, 1.0 mℓ of water and 1 mℓ of chloroform were added and again mixed. After centrifugation (1500 × *g*, 5 min) to resolve the phases, the heavier chloroform layer was evaporated to dryness *in vacuo*. The residue was taken up in 100 μℓ of methanol-chloroform (94:1) sonicated for 10 min and injected in 50 μℓ aliquots.[67]

A similar method entailing a hexane or heptane extraction of serum or red cells affords an extract that can be resolved by reverse phase HPLC.[68] The procedure not only permits resolution and quantitation of retinol and retinyl acetate, but tocopherols as well (see Table LC 18 and Section III.H.2.a). Greater sensitivity has been gained by using fluorimetric detection with direct and reverse phase HPLC.[69]

A method for the detection of all-*trans*- and 13-*cis*-retinoic acid and aromatic retinoic acid analogues in blood and urine by direct HPLC has been developed. This sensitive and specific chromatographic technique gave overall recoveries of 90 ± 5% and a limit of detection of retinoids of 10 to 20 ng/mℓ using 1 mℓ of blood.

To a 15 mℓ amber centrifuge tube (PTFE No. 16 stoppered) was added 1 mℓ of whole blood and 2.5 mℓ of buffer (1 *M* phosphate buffer, pH 7.0 for all *trans*-, 13-*cis* retinoic acid, Ro 10-9359 and Ro 10-1670 and 1 *M*-borate-Na_2CO_3-KCl, pH 9.0 for Ro 11-1430 analyses). The solution was mixed well with a Vortex mixer. To extract 13-*cis* retinoic acid and all-*trans* retinoic acid 6 mℓ of diethyl ether was used whereas for aromatic retinoic acid analogues Ro 10-9359, Ro 10-1670, and Ro 11-1430, ethyl acetate was the solvent of choice. The mixture was shaken for 15 min on a reciprocating shaker (Eberbach, Ann Arbor, Mich.) at 80 to 100 strokes per min. Together with the samples, control blood samples containing B_1, B_2, B_3, or B_4 (Table 1) were run for each assay, respectively. The samples were spun at 2500 rpm (1500 g) in a refrigerated centrifuge (Model PR-5, rotor 253; Damon/IEC Corp., Needham, Mass.) at 5°C for 10 min. A 5.0 mℓ aliquot of the organic (upper) layer was added to another amber 15 mℓ conical centrifuge tube and the solvent evaporated off in a N-EVAP evaporator (Organomation Associates, Worcester, Mass.) under a stream of nitrogen gas. The residues were taken up in 100 μℓ of the appropriate mobile phase and 1-μℓ aliquots were injected.[66] See Table LC 15-17 for details on the HPLC of A vitamers.

Aromatic retinoids in plasma have been assayed by extraction and separation by HPLC on silica gel columns with an online UV detector operated at 360 nm. The isocratic mobile

Table 1
PREPARATION OF STANDARD SOLUTIONS OF COMPOUNDS I-V FOR HPLC ANALYSIS

Standard solution	ng of compound per 0.1 mℓ of solution — Group A		Group B		Group C
	I	II	III	IV	V
B_1	25	25	50	100	25
B_2	50	50	100	200	50
B_3	75	75	150	300	75
B_4	100	100	200	400	100

I 13-*cis* retinoic acid
II all-*trans* retinoic acid
III aromatic retinoic acid analogue Ro 10-9359
IV aromatic retinoic acid analogue Ro 10-1670
V aromatic retinoic acid analogue Ro 11-1430

phases were mixtures of: (A) hexane-tetrahydrofuran-glacial acetic acid (98:1.5:0.6) and (B) hexane-methyl benzoate-propionic acid (87.5:12.5:0.35). The mixtures were boiled before use.

The prefractionation procedure was as follows. To a 20 mℓ amber centrifuge tube was added 0.5 mℓ of plasma, 50 μl of the internal standard solution (containing 50 ng of retinoic acid), and 4.5 mℓ of sodium citrate buffer, pH 6 (Titrisol; Merck). Samples prepared in this manner were stable for 1 week at −20°C. The solution was extracted with 10 mℓ of hexane with shaking for 10 min on a reciprocating shaker (Heidolph) at 60 rpm. Control plasma (0.5 mℓ) containing 50 mg of Ro 10-9359 or Ro 10-1670 were also processed. After centrifugation at 2000 g for 4 min at room temperature to separate the phases, approximately 9 mℓ of the organic (upper) layer was transferred into another amber 20 mℓ conical centrifuge tube. The organic layer was evaporated to dryness at 40°C with nitrogen gas. The residues were cooled to 4°C and taken up in two 100 μℓ aliquots of the mobile phase and filtered through a small swab of cotton wool into a microtube. A 100 μℓ aliquot was injected. Samples not measured immediately could be stored without alteration for 3 months at −20°C. With A as the eluting solvent and a flow rate of 1.2 mℓ/min, retention times of Ro 10-9359, Ro 10-1670, and the internal standard were 2.6, 6.0, and 4.1 min, respectively. However, with this solvent, a metabolite of Ro 10-9359 known to be found in plasma after multiple administration of the parent compound was not separated from the peak of the main metabolite, Ro 10-1670.

By using mobile phase B at a flow rate of 1.2 mℓ/min, retention times of Ro 10-9359, Ro 10-1670, and internal standard were 3.2, 8.1, and 4.2 min, respectively, and the new metabolite was separated sufficiently from the Ro 10-1670 peak and appeared 7 min after the injection.[70]

A comparable high pressure liquid chromatographic determination of 13-*cis*- and all-*trans*-retinoic acids in human plasma has been reported by Sporn and co-workers.[71,72]

Because of their physiological relationships, simultaneous assays for serum retinol and tocopherol, the latter another important fat-soluble vitamin, have been developed (see Section III.H.2.a).

b. Fluorimetric Analysis

Several fluorimetric determinations of A vitamers in blood have been developed to eliminate interference from carotenoids and other fluorescent constituents. In one approach[73] serum was prefractionated on microcolumns of silicic acid and interfering impurities eluted with hexane or light petroleum. Retinyl derivatives were then eluted with isopropanol. Fluorimetric analysis of the latter eluates correlated with the Neeld-Pearson modification of the Carr-Price colorimetric method ($r = 0.871$). This procedure was further refined to include a four-step elution from silicic acid using light petroleum, cyclohexane, cymene, and 12% acetic acid in methanol.[74] Cyclohexane eluted retinyl palmitate, cymene eluted retinol, and acetic-acid methanol eluted retinoic acids. This method gave a mean value for retinol of 24.5 ± 6.8 μg/100 mℓ plasma in 66 preschool children, which was 16 to 22% below the above-mentioned fluorimetric and colorimetric techniques. This drop may reflect the amount of other A vitamers present.

Other fluorimetric methods for A vitamers in blood have been developed to correct for phytoene and to determine its levels simultaneously.[75]

c. Colorimetric Analysis

Retinol and retinyl esters have been determined by batch separation on alumina followed by colorimetry with trifluoroacetic acid.[80] The drawbacks mentioned above for the Neeld-Pearson modification of the Carr-Price technique, for the methods using antimony trichloride, and for direct fluorimetric assays apply here as well. They all suffer from a lack of specificity and interference by carotenoids and related compounds. Comparisons of these methods with protocols utilizing HPLC illustrate this lack of specificity.[78]

3. Detection in Urine and Feces

Considerable effort has been devoted to the measurement of A vitamers or their analogues and their metabolites in mammalian urine after administration of pharmacological doses.

Initial attempts to isolate A vitamers entailed extraction with organic solvents followed by TLC. Sufficient amounts of catabolites were isolated to carry out spectral analyses. Recently, HPLC has also been exploited for the isolation and characterization of individual metabolites by mass and proton magnetic resonance spectrometry.[79]

Thirty-six rats were injected with doubly labeld retinoic acid (1.1 g in 40 mℓ of a 1:2 mixture of Tween 80-0.9% NaCl). Their urine was collected for a period of 5 days, adjusted to pH 2 with 2 *N* HCl, and prefractionated on Amberlite XAD-2 resin (75×6.5 cm). The column was washed with 1.2 ℓ of water and the unconjugated metabolites were removed with 2.4 ℓ of ethyl acetate. This fraction contained 69% of the urinary ^{3}H- and 61% of the ^{14}C activity and was purified further on a column (60×5.5 cm) of silica gel 60 (Merck, 0.04 to 0.063 mm) by gradient elution (pentane-dichloromethane-acetonitrile-methanol). This procedure gave three radioactive fractions that contained about 40% of the ^{3}H- and 10% of the ^{14}C activity. Fractionation by TLC with detection by UV radiation and radioactivity and by HPLC with UV detection gave three compounds: 5-methyl-5[2-(2,6,6-trimethyl-3-oxo-1-cyclohexen-1-yl)vinyl]-2-tetrahydrofuranone; 5-[2-(6-hydroxymethyl-2,6-dimethyl-3-oxo-1-cyclohexen-1-yl)vinyl]-5-methyl-2-tetrahydrofuranone; and 6-(6-hydroxylmethyl-2,6-dimethyl-3-oxo-1-cyclohexen-1-yl)-4-methyl-4-hexenoic acid. The carboxyl group of the latter was methylated by treatment with diazomethane in ether 3 min before HPLC analysis.

A similar protocol permitted the detection of fecal metabolites of retinoic acid.[80]

Methods exploiting HPLC that were developed for the detection of all *trans* and 13-*cis* retinoic acid in plasma (see Section III.G.2.a) have also been used for the determination of physiological amounts in urine. Radiometric measurements of retinoic acid metabolites in urine and plasma have also been made with the aid of HPLC. Normal and reverse-phase chromatography were used.[75]

4. Detection in Tissue

The determination of A vitamers in mammalian tissues is complicated by the fact that they constitute a minute fraction of the total lipid pool. Large scale separation and purification of synthetic A vitamers originally included column chromatography on alumina or less frequently, silicic acid, silica gel, dicalcium phosphate, as supports for partition and ion exchange chromatography. However, the lability of vitamin A compounds resulted in poor recoveries, incomplete resolution, and artifact formation in attempts to study their metabolism using the above methods.[82]

Retinol, retinal, and retinoic acid have been resolved by liquid-gel partition chromatography on Sephadex LH-20 with solvent mixtures of chloroform, Skellysolve B and methanol. Retinyl ester, retinol, retinal, and retinoic acid have been separated on hydroxyalkoxypropyl Sephadex using Skellysolve B and acetone. These protocols are suitable for preparative work, as well as for metabolic studies, because of the lack of decomposition and complete recoveries experienced.[83] Vitamin A ester metabolism was also investigated by administration of labeled retinyl acetate to animals, followed by sacrificing and tissue removal. After addition of "cold" carrier to the tissue extracts, liquid partition chromatography provided excellent resolution of the metabolites, which were counted and their purity determined, by UV spectroscopy. This approach also permits measurement of long chain fatty acid esters such as retinyl palmitate.[83]

An even better separation of retinyl esters, as well as related retinols, can be achieved by reverse-phase HPLC. The speed and sensitivity of this method, when used with UV detectors, obviates the use of radioisotopes and permits direct isolation and detection.

After the tissues (1 to 10 g) were homogenized with 2 parts water containing 0.5 mg of ascorbic acid and 0.5 mg EDTA in a blender, the homogenate was lyophilized. Ether containing 0.1% butylated hydroxytoluene was used for extraction. An aliquot of the extract was evaporated to dryness under a stream of nitrogen, and the residue was redissolved in 100% chloroform, followed by the addition of 0.5 volume of methanol to give a final composition of 2:1. This solution was centrifuged to clarity and analyzed directly by HPLC. All procedures were carried out in minimal light or under red light, and sample solutions were stored in the dark at −20°C prior to chromatography. See Table LC 15 for conditions of separation by HPLC.[78]

Retinol pigment epithelium A vitamers, partially purified by sucrose density gradient centrifugation to enrich them in rod outer segments, have been subjected to direct HPLC to resolve isomeric retinyl esters and other visual pigments. See Table LC 17.[84]

5. Phosphorylated and Glycosylated Retinols

It has been learned that mammalian tissues synthesize retinyl phosphates, retinyl phosphate glycosides, and retinyl glycosides.[85] Some of these reactions are reminiscent of transformations observed for polyprenols (see Section III.E.2). Labeled retinol has been incorporated into hepatic mannolipid fractions in vivo and in vitro. The fractions were purified by chromatography on DEAE-cellulose and silica gel. The mannolipids were then hydrogenated, and after removal of water-soluble compounds, the lipid phase, which contained the label from the retinol, was separated on inactivated alumina columns.

The mannolipid mixture was resolved by TLC or lipid-partitioning into dolichol and retinyl derivatives. Retinyl phosphate and retinyl phosphate mannose have also been separated on silicic acid and DEAE-cellulose acetate columns. Recoveries were poor (40% or less), and if the silicic acid is not treated with ammonia, retinyl phosphate is converted to anhydroretinol. Chromatography on Sephadex LH-20 also causes this displacement of the phosphate moiety. Recently, reverse-phase HPLC has been used to resolve retinyl phosphate, mannosyl retinyl phosphate, retinoic acid, retinoic acid esters, retinal, and retinol (Figure 10).[86] In studies of the in vitro metabolism of labeled retinol, the phosphorylated derivatives were

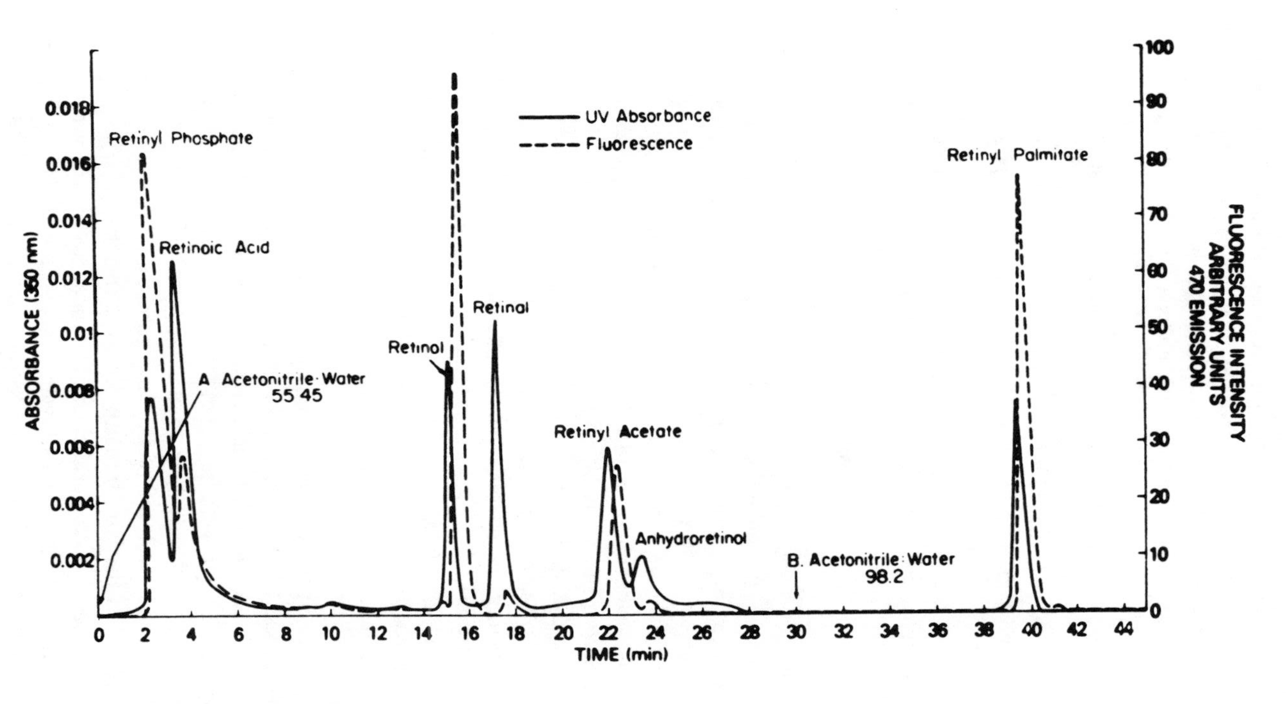

FIGURE 10. Reverse-phase HPLC of standard A vitamers.

Instrument: Altex 322 MP programmable LC system with a Hitachi Model 100-30 variable wavelength (195-850 nm) spectrophotometric detector and a Schoeffel Fluorimeter Model FS970.

Column: Whatman Partisil-10-ODS (25 cm × 4.6 mm I.D.).

Solvents, flow rates: Acetonitrile-water (55:45) at an initial flow rate of 1.0 mℓ/min for the first 10 min, 3.5 mℓ/min for the following 20 min; Acetonitrile-water (98:2) at a flow rate of 1.2 mℓ/min for the final 15 min.[86]

added as carrier to intestinal extracts, and 90 to 95% were recovered after HPLC. In this manner, the extent of phosphorylation and mannosylation (12 to 42%) and esterification (19 to 23%) of retinol was assessed radiometrically. Similarly doubly labeled retinyl phosphate galactose has been isolated from galactolipid extracts by DEAE-cellulose and silica gel chromatography in tracer experiments with [^{3}H]-retinol and [^{14}C]-galactose.

Retinyl glycosides have been separated from retinol on cellulose columns eluted with *n*-butanol-water (78:17). Descending chromatography on Whatman No. 3 paper has also been utilized for the identification of glycosides. The solvent system consisted of *n*-butanol-acetic acid-water (78:5:17). All procedures were performed in subdued light in order to prevent isomerization of retinol.[87]

6. Oxyretinoic Acids

Isolation of rat intestinal 5,8-oxyretinoic acid has been accomplished by a combination of liquid-partition, ion exchange and HPLC an excellent demonstration of the power of resolution and sensitivity of modern chromatographic techniques.

All steps of the purification of the retinoic acid metabolites were performed under red light and the isolated compounds were stored at −70°C in degassed solvents. All-*trans* retinoic acid was added to [11,12-^{3}H]-retinoic acid (specific activity, 11.1 Ci/mmol) to yield a final specific activity of 8.1 × 10^5 dpm/μg. Each of 63 vitamin A-deficient rats were administered 450 μg of the [^{3}H]-retinoic acid 3.5 hr prior to killing. The entire small intestine was excised and washed with cold 0.9% NaCl. The intestinal mucosa (126 g) was scraped off and homogenized in 1 vol of an aqueous solution of ethylenediaminetetracetic acid (EDTA) and propyl gallate (50 μg/mℓ). After lyophilization the residue was extracted wtih 1.5 ℓ of chloroform-methanol as described by Ito et al.[83] The chloroform-methanol was evaporated and the residue was taken up in methanol. All of the radioactivity present in the intestine was recovered in this methanol-soluble fraction. After removal of the methanol, the residue was partitioned between 0.2 ℓ of ether and 1 ℓ of 2% aqueous NaOH. The aqueous phase was back-extracted with 4 × 1 ℓ of ether and the combined organic phase designated Ether 1. The aqueous phase was brought to pH 1 with 6 *N* HCl and extracted with 5 × 250 mℓ of ether. This organic phase (Ether 2) contained the peak 8 metabolites. After evaporation, the residue from Ether 2 was partitioned between methanol and hexane (250 mℓ). The methanol fraction (MeOH 2) was back-extracted with 4 × 250 mℓ of hexane. Of the total intestinal mucosal radioactivity, 25% or 250 μg of vitamin A metabolites as determined from the specific activity of the administered [^{3}H]-retinoic acid was present in this methanol extract (MeOH 2).

A fraction of the methanol extract (60% of the radioactivity, 140 μg of A vitamers) was applied to a 2 × 55 cm Sephadex LH-20 column and eluted with acetone (column 1). A single radioactive peak (132 mg) was obtained and fractionated on a 1 × 148 cm column of Sephadex LH-20 (column 2). Elution with methanol afforded 117 μg which was added to a 1 × 20 cm column of DEAE-Sephadex A-25 (hydroxide form). Column 3 was eluted with 100 mℓ of methanol followed by 100 mℓ of methanol-formic acid (99:1). The peak 8 compounds (101 μg) were eluted in the last 50 mℓ of acidified methanol.

A fourth step entailed reverse-phase HPLC of the peak 8 compounds (33 μg) which were eluted with 0.01 *M* ammonium acetate in methanol-water (60:40) to give 34% of the total radioactivity (11.6 μg) as component 8_{II}. The latter was subjected to straight-phase HPLC (column 5) with tetrahydrofuran-hexane-formic acid (7:93:0.1) as solvent. The peak fraction (9 μg of 8_{II}) was methylated with diazomethane and chromatographed with a solvent system of tetrahydrofuran-hexane (1:99). Methyl 8_{II} (7 μg) were recovered by this protocol.

The metabolite 8_{II} was identical with 5,8-oxoretinoic acid by UV and mass spectral and chromatographic comparison with an authentic sample.[88] In vitro epoxidation of all-*trans* retinoic acid has been assayed radiometrically with the separation of product 5,6-epoxide by reverse-phase HPLC.[89]

A radioimmunoassay for purified retinol has been developed by the preparation of antibodies of retinoic acid conjugated to bovine serum albumin. Application to tissue determinations was complicated by the water insolubility of the lipids in the aqueous system used for immunotitration.[90]

7. Isomeric A Vitamers

Synthetic geometrical isomers of retinoids are readily resolved by HPLC. All-*trans*, 9-*cis*, 11-*cis* and 13-*cis* retinols and retinyl palmitate have been resolved simultaneously on silica gel columns (see Tables LC 16 and 17).[91] The 7-*cis* isomer of retinol has also been resolved from the other mono *cis* isomers of the aldehydo-retinoids by elution of a silica column with light petroleum-diethyl ether (92:8).[92] Reverse-phase HPLC on octadecyl-silica has been used to simultaneously separate all-*trans*, 9,11,13-tri-*cis*, 11,13-di-*cis*, 13-*cis*, 11-*cis* and 9-*cis* methyl retinoates. With methanol-water (7:3) as eluting solvent, seven of the isomeric retinoic acids were resolved.[93] Straight and reverse-phase high pressure liquid chromatographic separation of a broad range of oxygenated retinoids has been accomplished. The reverse-phase system was also used to resolve all-*trans* and 13-*cis* retinols (Figure 11).[94]

8. Gas-Liquid Chromatography of Hydrogenated Retinols

Like the carotenoids, A vitamers are thermally labile and separation by GC has only been accomplished after their hydrogenation. The method has been evaluated for quality control of retinol and retinol acetate. In each case the hydrocarbon is a side product of the hydrogenation reaction, and for quantitation it must be included in peak height analyses. The protocol adopted is as follows. A solution of 25 mg of retinol in 96% ethanol was transferred to a 15 mℓ vacutainer tube. Five to ten milligrams of PtO_2 were added and the volume made up to 2 mℓ with 96% ethanol. A 1.5 in. 20 G needle attached to a three-way stopcock was inserted through the vacutainer stopper. The vacutainer was evacuated twice and the air replaced with hydrogen gas at a pressure of 5 to 10 psi. The vacutainer tube was shaken until the reaction mixture was colorless (approximately 5 min) and then allowed to stand at room temperature overnight. Methylheptadecanoate (4.2 mg) in 0.2 mℓ of 96% ethanol was added as an internal standard. After thorough mixing, the PtO_2 was allowed to settle and 1.6 μℓ of the supernatant containing the hydrogenation products was analyzed by GC in a Bendix GC-2500 chromatograph equipped with a flame ionization detector. A 6-ft glass column ($^1/_4$ in. O.D.), packed with 5% OV-1 on acid-washed and DMCS-treated 80-100 mesh Chromosorb W was operated isothermally at 230°C. The inlet and transfer zone temperatures were 240°C. Helium was used as the carrier gas at a flow rate of 40 mℓ/min.[95]

H. ISOPRENOID QUINONES

1. General Aspects

The isoprenoid quinones are comprised of three main families. The first group, which includes plastoquinones and the E vitamers (tocopherols), are derivatives of methylated benzoquinones. The second are the ubiquinones, which possess a 3,4-dimethoxy toluquinone unit. The final family are the menaquinones, which contain a monomethylated naphthaquinone moiety and encompass the K vitamers. Attached to all of these quinones are single polyprenyl side chains of various length depending upon the species. The isoprenoid chain, often a phytyl group in plants, e.g., tocopherols and phylloquinones, can contain up to 13 isoprene residues in bacteria and mammals, thereby imparting to the molecule chromatographic behavior reminiscent of the polyprenols. In fact methods developed to isolate the latter afford the isoprenoid quinones as well, and the polarity of the substituted aromatic moiety permits ready resolution by adsorption chromatography using alumina or silica gel (see Section III.E.2). Like the polyprenols, the quinone polyprenyl side chain is often found

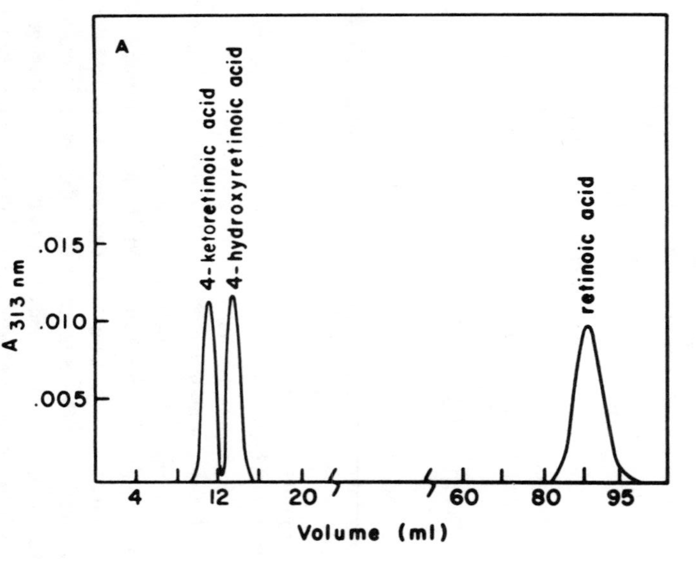

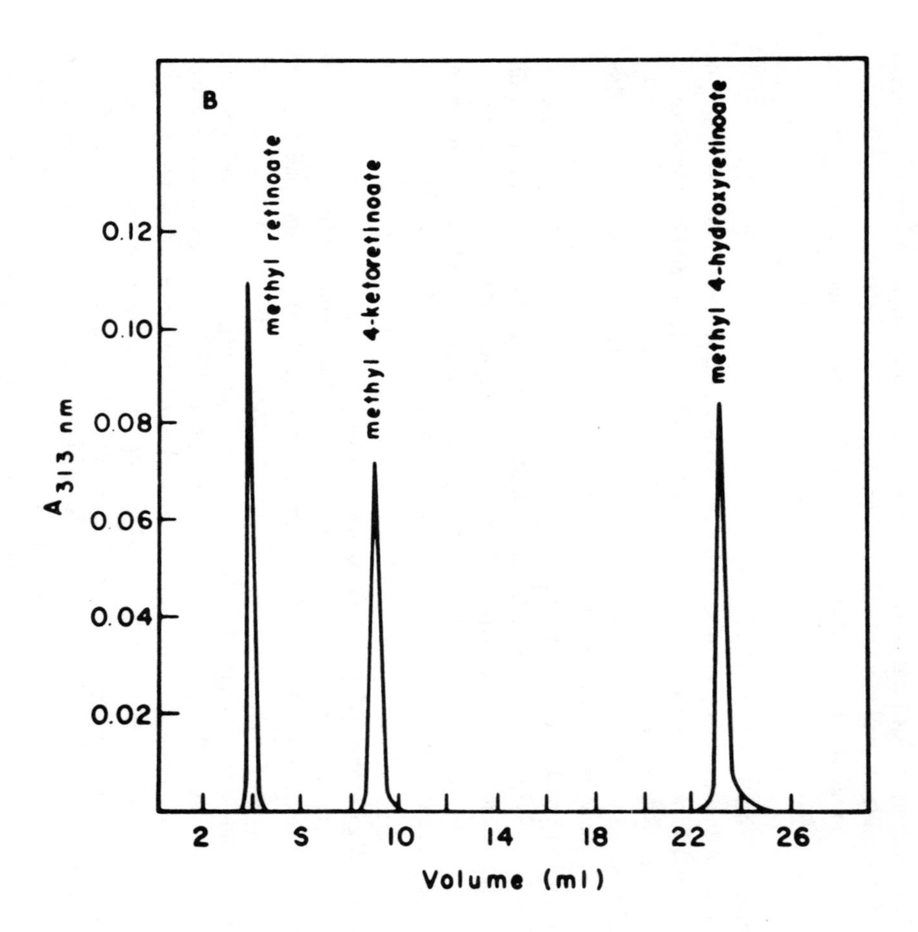

FIGURE 11. Reverse-phase (A) and straight-phase (B) HPLC of polar A Vitamers.

Instrument: Waters Associates Model ALC/GPC-204 equipped with a Model 6000A solvent delivery system with a fixed wavelength (313 nm) UV detector.

Columns: μBondapak C_{18} (30 cm × 4 mm I.D.) in stainless steel for reverse-phase; μPorasil (30 cm × 4 mm I.D.) in stainless steel for straight-phase.

Solvent, flow rate: 0.01 *M* Ammonium acetate in methanol-water (65:35) for reverse-phase; tetrahydrofuran-hexane (7.5:92.5) for straight-phase. All columns were eluted at 2 mℓ/min.[94]

with several of its double bonds reduced. In the case of the tocopherols, chromanol forms in which the quinone is reduced to a monohydroxyalkyl ether (formed with the 3′ carbon of the isoprenoid side chain) are prevalent in mammals and plants. Normally, simple dihydroquinones are readily oxidized in air back to the quinone form. Oxygen can cause degradation of these quinones in the presence of base, so that sample preparation requiring saponification must be carried out under anaerobic conditions. Pyrogallol or ascorbate are routinely included in the base mixture. It is also advisable to include antioxidants such as 2,6-ditertiary-butyl-*p*-cresol (BHT) in extracting solvents.

Like the other fat-soluble vitamins, the E and K vitamers are trace constituents in mammalian fluids and tissues. Reasonably accurate and specific assays other than bioassay have only recently been developed for these vitamers. Ubiquinones are found in higher concentrations as mitochondrial electron transport components in mammalian tissue.

2. Detection in Plasma and Other Body Fluids

Of the terpenoid quinones only the E vitamers occur in blood in sufficient amounts (3 to 15 μg/mℓ of serum) to be detectable by modern analytical methods. (Following pharmacological doses of vitamin K_1, GC has been used to detect as little as 3 ng/mℓ in plasma, see later.) Serum tocopherols are associated with the β-lipoprotein fraction, the size of which will influence the concentration of vitamin E. Until recently, colorimetry, fluorimetry, TLC, and GC have been used to monitor serum levels of the E vitamers.[96] As in the case of retinoids, however, colorimetric and fluorimetric methods suffer from interfering substances, such as carotenoids, and neither are capable of distinguishing between the various vitamers. In fact d-α-tocopherol (the 5,7,8-trimethyltocol) represents only 88% of the total E pool in serum, while d-β-(5,8-dimethyltocol, 2%) and d-γ-(7,8-dimethyltocol, 10%) derivatives comprise the remainder. Both GC and TLC methods suffer from interference by other lipids (particularly cholesterol in the former) and require extensive prefractionation. High temperatures are required for GC analysis. Thermal lability, and the laborious and time-consuming nature of that procedure, also deters from overall usefulness. On the other hand, HPLC offers a rapid, sensitive, and simple means to analyze tocopherols.

a. HPLC

Both direct and reverse-phase HPLC with UV detection have been utilized in the separation of plasma tocopherols. Essentially the same sample preparations of deproteination and *n*-hexane extraction of plasma preceded the HPLC.

After addition of an internal standard (2 μg of tocol in 4 μℓ of methanol) to serum (200 μℓ), it was deproteinized by treatment with 0.2 mℓ of ethanol and extracted with 0.5 mℓ of *n*-hexane, mixing for 4 min in a Vortex mixer. The mixture is centrifuged at 3000 rpm for 5 min and the supernatant was transferred to a 10 × 100 mm conical tube. The aqueous layer was re-extracted with 0.5 mℓ of *n*-hexane and the combined organic layers evaporated to dryness by heating at 40°C in a stream of dry nitrogen. The residue was taken up into 200 μℓ of methanol and half this volume was injected onto an RP 18 (octadecyl silica) HPLC column and eluted with methanol.[97]

Bieri and colleagues[68] have developed a similar method for simultaneous determination of α-tocopherol and retinol in plasma by hexane extraction without prior deproteination. Reverse phase HPLC on a μBondapak column resolved most of the tocopheryl derivatives, retinol, and retinyl acetate (see Table LC 18). The same method was successfully applied to the analysis of red cells. In this case prefractionation entailed washing the red cells three times with 0.9% saline and preparing a 50% suspension in saline containing 0.5% pyrogallol. A hematocrit value was obtained immediately. A 0.5 mℓ aliquot was then transferred to a 12 × 100 mm screw cap (or glass stoppered) test tube to which 1.5 mℓ of cold methanol (cooled in dry ice-acetone bath) was then added drop-wise with slow vibration on a Vortex mixer. Protein precipitated as a fine suspension without clumping. Thereupon 20 μℓ of tocopheryl acetate (40 to 50 μg/mℓ) was added together with 2 mℓ of heptane. (Since 1 mℓ of this solvent is taken to dryness, any contaminants will be amplified in the HPLC chromatogram. Heptane was superior to hexane for this reason.) After mixing the tubes on a Vortex mixer for 45 sec, the phases were separated by centrifugation. Approximately 1 mℓ was transferred to a 5 mℓ centrifuge tube, with care taken to avoid removing any protein. After evaporation of the heptane with a stream of nitrogen in a 60°C water bath, the mixture was dissolved in 25 μℓ of diethyl ether followed by 75 μℓ of methanol. About 90 μℓ of this solution is injected, using a 10 μℓ methanol flush in the syringe.[68]

For direct HPLC of tocopherols, a Corasil column was eluted with *n*-hexane-disopropyl ether (96:4) at 60 mℓ/hr.[98] Neither these reverse-phase nor the direct HPLC methods were successful in resolving the positional isomers β- and γ-tocopherol, but they all gave excellent recoveries (89 to 100%), low standard deviations, and excellent within-run precision (2.3% with reverse phase HPLC). The eluted tocopherols in the latter system were subjected to GC-MS to confirm the identity of the serum constituents as E vitamers.[97]

Using essentially the same sample preparations, but a column of heavily loaded octadecyl bonded phase (18% organic material), a reverse-phase HPLC method for the simultaneous determination of retinol and α-tocopherol has been developed. Methanol at 2 mℓ/min was used to elute the fat-soluble vitamins.[99]

Reverse-phase HPLC with fluorescence detection (excitation at 296 nm and emission at 330 nm) affords the greater sensitivity required for the analysis of plasma tocopherols in domestic animals. Since animal tocopherol plasma levels are an order of magnitude lower than human concentrations, UV detection is not suitable for the former, especially if a deficiency in this vitamin exists.[100]

Earlier spectrofluorimetric methods for both serum, plant, oil, and tissue tocopherols have included a low pressure LC system.[101] Isomers of tocopherol and tocotrienols were resolved on hydroxyalkoxypropyl Sephadex columns eluted with hexane-benzene-ethyl acetate mixtures.

b. GC and TLC

Before the advent of HPLC, Bieri and co-workers developed a method for determination of α-tocopherol in erythrocytes, which entails a prefractionation by TLC and subsequent GC.[102] Labeled α-tocopherol was included as an internal standard. No attempts were made to determine other tocopherols. The procedure used an initial hexane extraction because this solvent removed less cholesterol, which, if present in large amounts, distorted the α-tocopherol-quinone GC peak, especially when SE-30 was used as column packing. The latter is produced by ferric chloride oxidation from α-tocopherol in this procedure. Since the HPLC-based methods described above (Section III.H.3.a) are simpler, more rapid and more accurate, this GC procedure is no longer preferred for routine analysis.

A serum α-tocopherol analysis entailing saponification, TLC, and GC with solid injection has been developed. Recoveries were 70 to 80% and reproducibility was good, even for low serum levels.[103]

A similar method, which concludes with GC of trimethyl silyl ethers of tocopherol, is also available.[104] This method affords a determination of both α and γ isomers. Lovelady has reported a modified procedure that allows detection of all four isomeric tocopherols in plasma and erythrocytes. The procedure is as follows. Plasma (3 mℓ) and red blood cells (3 mℓ) were deproteinated by addition of 6 mℓ of 95% ethanol to each in separate 15 × 150 mm glass-stoppered test tubes and heating on a block at 50°C for 25 min. The tubes were then removed, allowed to cool and treated with 1.5 mℓ of 10 *N* NaOH and heated for an additional 40 min. After removal from the block, 3.0 mℓ of water was added and each tube mixed vigorously. After extraction of the solutions with 12.0 mℓ of *n*-hexane (spectroanalyzed grade), the organic phases were transferred to separate glass-stoppered centrifuge tubes and evaporated to dryness in a stream of nitrogen. The residues were taken up in 1.0 mℓ of *n*-hexane for TLC analysis (see also Table TLC 36).

Aliquots of the hexane extract were applied to silica gel G plates and resolved with cyclohexane-*n*-hexane-isopropyl ether-ammonium hydroxide (conc.) (40:40:20:2). Development was performed in closed rectangular tanks (24 × 23 × 8 cm) lined with 3 mm filter paper and presaturated for 30 min with the mobile phase. The solvent ascended 17 cm on the plate in about 90 min. After chromatogram development, the plates were dried and sprayed with phosphomolybdic acid - dichlorofluorescein (see Section II) until uniformly yellow. Drying the plates with an air jet dryer for 3 to 4 min led to the development of color in the tocopherol bands. They were detected as fluorescent bands under 2540 Å UV light.

The tocopherol bands were scraped from the plate, transferred to 15 mℓ glass-stoppered centrifuge test tubes, and extracted with 1.0 mℓ of methanol. After centrifugation to remove the silica gel, the methanol was transferred to 15 mℓ glass-stoppered centrifuge tubes and evaporated to dryness in a stream of nitrogen gas.

The residues were taken up in 0.2 mℓ of *n*-hexane and 0.25 mℓ of bis-(trimethylsilyl)trifluoroacetamide and 0.05 mℓ of bis(trimethylsilyl)ether acetamide-trimethylchlorosilane (5:1) were added. After vigorous mixing the samples were heated at 50°C in a block for 45 min and evaporated to dryness with nitrogen. The residues were dissolved in 0.25 to 1 mℓ of *n*-hexane for use in GC.

GC was performed with a Beckman GC-5 chromatograph equipped with a dual hydrogen flame detector and a 6 ft × $^1/_3$ in. O.D. stainless steel column packed with 3% OV-1 on 80 to 100 mesh Supelcoport (Perco Supplies, San Gabriel, Calif.). After preconditioning the column for 24 hr at 250°C, the column was operated isothermally at 240°C. The detector was maintained at 290°C, carrier gas helium flow rate was 80 mℓ/min at 52 psi, hydrogen pressure was 40 psi, and air pressure was 50 psi.[105]

All of the TLC-GC methods are lengthy, but a rapid capillary GC method for the detection of α-tocopherol and its isomers and free fatty acids has been reported by Lin and Horning.[106] The procedure is as follows. To a plasma sample (1 mℓ) in a 15 mℓ tube with a PTFE-lined screw cap, containing 30 to 50 μg of heptadecanoic acid in 30 to 50 μℓ of pyridine as internal standard, was added 5 mℓ of isopropanol-isooctane-1 *N* sulfuric acid (40:10:1) solution. The tube was vigorously mixed (Vortex mixer), 2 mℓ of water and 5 mℓ of isooctane were added and the mixing repeated. The phases were separated by centrifugation and the organic layer was evaporated to dryness with a stream of nitrogen gas.

The acidic components were converted to methyl esters by a short treatment of the lipid sample dissolved in 0.5 mℓ of methanol with diazomethane (2 mℓ prepared from Diazald, Aldrich, Milwaukee, Wis.). Solvents and excess reagent were removed with nitrogen gas after 10 to 15 min and the residue was treated with 50 μℓ of pyridine, 50 μℓ of bis (trimethylsilyl)acetamide and 25 μℓ of trimethyl-chlorosilane. After heating the mixture for 1 hr at 90°C, it was used directly for GC.

Analyses were performed with an LKB 9000 gas chromatograph-mass spectrometer equipped with a 4 m × 3.4 mm I.D. glass coil containing 3% PZ-176 on 80 to 100 mesh acid-washed and silanized Gas Chrom P. The conditions were similar to those used for long chain methyl esters and sterols with temperature programming at 2°C/min from 180°C. Mass spectra obtained were identical to those of authentic samples. The tocopherols appear after 40 min and cholesterol does not interfere with their elution under these conditions.[106]

TLC has been utilized in a procedure that begins with the saponification of serum cortisol. The α-tocopherol was quantitated by densitometry. The protocol is an adaptation of Bieri's method[107] wherein ethanol was added and after saponification hexane was used to extract the tocopherols. The procedure is as follows. Tissue (0.5 g) was homogenized in 2 mℓ of 2% pyrogallol in ethanol (freshly prepared) and after addition of 0.5 mℓ of 60% NaOH, it was heated under purified nitrogen gas at 70°C in a stoppered tube for 20 to 25 min with occasional shaking. To reduce losses due to light and air oxidation, the nitrogen was scrubbed with Fieser's solution to remove O_2, and these steps were performed under low intensity lighting.

After cooling the solution, an equal volume of water was added, and it was mixed vigorously with 2 × 3 mℓ of 0.00125% BHT in hexane for 2 min. The organic layers were combined and dried at 50°C with a stream of nitrogen and the residue taken up in benzene and added to activated silica gel TLC plates. No more than 35 mg was applied to a region on the plate <5 mm in diameter. Benzene-ethanol (99:1) was the developing solvent (α-tocopherol R_F = 0.21 ± 0.021) unless quinols were present whereupon it was necessary to develop the plate in a second dimension with hexane-ethanol (9:1) (α-tocopherol R_F = 0.37 ± 0.063). Development was performed in the dark.

After heating the plates for 18 to 20 hr at 110 to 112°C, plates were cooled and scanned with a Schoeffel SD 3000 densitometer (Schoeffel Instrument Corporation, Westwood, N.J.) in the transmission mode using a 1 × 5 mm slit with the monochromator slit at 1.5 nm and the wavelength at 270 nm. Samples containing less than 10 μg of α-tocopherol were calibrated to 0.4 as full scale absorbance. The reference beam was focused on a part of the plate devoid of sample. A 20 × 20 cm plate was used for 12 samples. Peak height and area correlated well with standards using a recorder integrator that measured the area from peak to base line.[108]

Other TLC systems have been used for serum samples (see Table TLC 36).[109]

Although the K vitamers are not detectable in normal serum, methods are available to detect K_1 and its 2,3-epoxide in plasma by electron capture GC after administration of pharmacological doses. The procedure is as follows. To plasma samples (0.2 to 0.8 mℓ) in 15 mℓ PTFE-lined screw-cap glass tubes containing 90 μℓ of an ethanol solution of vitamin K_2 (1.5 μg/mℓ) as internal standard were added 2 mℓ of double distilled water and 10 mℓ of *n*-hexane-ethanol (1:1). After 30 min on a rotary mixer at 25 rpm, the phases were separated by centrifugation and the organic layer dried in a conical tube with a stream of nitrogen. The residue was taken up in 10 to 25 μℓ of *n*-hexane and 1 to 2 μℓ aliquots injected into the gas chromatograph. Care was taken to remove lipids from glassware by prewashing with hexane. Precaution was also taken to avoid destruction of the sample by light.

Analyses were performed with a DANI gas chromatograph (Model 3600) equipped with an electron-capture detector (radioactive source ^{63}Ni, 10 mCi, operated in a pulse mode with modulated frequency) and a 190 cm × 2.2 mm I.D. O-shaped silanized pyrex glass column containing 3% OV-17 on 90-100 mesh Anakrom Q (NEN Chemicals, Dreieichenhain, G.F.R.) The instrument was operated with a column oven temperatures of 302°C, injection port temperature of 315°C, detector temperature of 305°C, and a flow rate of 80 mℓ/min for the oxygen-free nitrogen carrier gas. A lower limit of detection was estimated to be 3 ng/mℓ plasma.[110]

c. Fluorimetry and Spectrophotometry

The fluorimetric detection and estimation of E vitamers had been developed as early as 1942,[111] and since then it has been applied to serum analysis.[112] As discussed in earlier sections, fluorimetry has drawbacks in distinguishing between vitamers and being susceptible to considerable interference. A recent procedure permits simultaneous assay of vitamin A and E in serum and plasma.[113] The method entails a hexane extraction and fluorimetry (excitation at 295 nm and emission at 320 nm). It tends to give lower values than the colorimetric method, but it still represents an overestimation as shown by prefractionation on columns of hydroxy-alkoxypropyl Sephadex with hexane-benzene-ethyl acetate (67:30:3) as the eluting solvent. The latter chromatography demonstrates that β- and γ-tocopherols contribute to the overestimation of α-tocopherol, but apparently other lipids also interfere. A more elaborate spectrophotometric method for the determination of E vitamers in erythrocytes and plasma includes saponification, and TLC to remove interfering lipids. It is more rapid than GC techniques. The procedure is as follows. One volume of 1% EDTA was added to 50 volumes of fresh venous whole blood in a plastic round-bottom centrifuge tube and mixed by gently inverting the tube 3 to 4 times. The red blood cells were sedimented by centrifugation at 5000 g and the plasma and buffy layer removed by aspiration. The red blood cells were washed with 3 × 5 volumes of an isotonic phosphate-buffered saline solution containing EDTA, pH 7.4. The buffer contained 1.42 g of anhydrous Na_2HPO_4, 7.27 g of NaCl, and 0.1 g of Na_2 EDTA in a final volume of 1 ℓ. After the third washing the final hematocrit of the red blood cells were made up to about 50% and measured accurately in a standard Adams microhematocrit centrifuge.

The red blood cells (2 mℓ) were slowly combined with mixing to 10 mℓ of 2% alcoholic pyrogallol in a 50 mℓ glass-stoppered centrifuge tube. After thorough mixing, the loosely stoppered tubes were heated in a water bath at 70°C for 2 min. After removing the tubes from the bath, 0.5 mℓ of saturated aqueous KOH was added and they were returned to the 70°C bath for an additional 30 min. Thereupon they were cooled in an ice bath and after addition of 7.5 mℓ of water and 22 mℓ of hexane, the tubes were stoppered and shaken vigorously for 2 min. The phases were separated by centrifugation at 1500 rpm.

The organic (upper) layer (20 mℓ) was transferred to a conical tube and dried with a stream of nitrogen. The residue was taken up in chloroform (20 to 50 μℓ) and the solution applied to silica gel G plates. Standard α-tocopherol was co-developed in a solvent of benzene-ethyl acetate (2:1) or benzene-ethanol (99:1). The tocopherol band was localized by spraying with 0.001% methanolic Rhodamine-5G and examining the plate with UV light. Prolonged exposure to UV light will destroy the tocopherol. The area of silica gel corresponding to the R_F of tocopherol standard was scraped from the plate and mixed with 1.5 mℓ of absolute ethanol with a vortex mixer. The silica gel was removed by centrifugation at 2500 rpm for 5 min.

To each of a series of 4 mℓ glass test tubes was added 1 mℓ of the unknown ethanol solution or standards of 1, 2, 5, or 10 μg of α-tocopherol in 1 mℓ of absolute ethanol. To each tube was added 0.2 mℓ of 0.2% ethanolic bathophenanthroline with mixing (Vortex mixer). From this point on, the subsequent steps were performed rapidly with care taken to avoid exposing the solution to direct light. Next 0.2 mℓ of an ethanolic solution of 1 m*M* $FeCl_3$ was added with mixing. After 1 min 0.2 mℓ of 40 m*M* phosphoric acid in ethanol was added and upon mixing, the absorbance of the solution, was measured at 534 nm using an automatic sampler and a Gilford model 240 spectrophotometer.

The tocopherol levels in plasma were also determined using the above protocol. Saponification was performed as before except 1 mℓ of plasma was treated with 2 mℓ of the 2% ethanolic solution of pyrogallol and after 2 min of heating at 70°C, 0.3 mℓ of a saturated KOH was added. The tubes were again heated for 30 min at 70°C, cooled in ice and treated with 1 mℓ of water and 4 mℓ of hexane. After 2 min of shaking, the tubes were centrifuged to separate the hexane and 3 mℓ of the phase was treated as above. The method measures total tocopherol since the TLC does not resolve α-tocopherol from its β-, γ-, or δ-derivatives. The average recovery of tocopherol from plasma was 89% and from erythrocytes, 70%.[114]

TLC was used to detect α-, β-, γ-, and δ-tocopherols in human milk. The protocol included extraction with ethanol, ethyl ether, and petroleum ether, followed by saponification in the presence of pyrogallol. Nonsaponifiables were then dissolved in methanol-ether (2:3) and passed through alumina to remove free fatty acids. The extract was next dissolved in benzene and passed through a Florisil column, and the eluate was used for TLC and GC-MS. After developing the mixture on silica gel plates with petroleum ether (b.p. 60 to 80°C)-isopropyl ether-acetone (85:12:4), zones were scraped off, extracted with absolute ethanol, and the silica gel removed by centrifugation before GC-MS. The latter was carried out on a 2% OV-17 stationary phase.[115]

Urinary metabolites of K vitamers were first detected using radiometric methods, but preparative isolation protocols of their glucuronides have been developed.[116,117] In this isolation procedure, urine at pH 4 was passed through an Amberlite XAD-2 resin and acidic conjugates were then subjected to continuous liquid-liquid ether extraction at pH 2. After β-glucuronidase treatment, aglycones were extracted with chloroform and subjected to reverse-phase TLC (see Tables TLC 33—35). For acidic aglycones, methylation with diazomethane preceded the TLC. Final purification was accomplished by adsorption TLC.

3. Isolation from Tissues

a. HPLC

The simplicity, speed, and sensitivity of HPLC for the isolation of E vitamers is evident from recent investigations. Van Niekerk was able to inject plant oils directly onto silica gel columns (see Table LC 19) that resolved all four isomers. A general method for both oils and solid tissue using a simple extraction technique for prefractionation has also been developed. Eight isomeric tocopherols and tocotrienols were resolved by direct HPLC on silica gel with spectrofluorimetric detection.

A 100 mℓ isopropanol solution of the sample (10 g) was boiled for 10 min and homogenized for 1 min in a Virtis homogenizer. After addition of 50 mℓ of acetone, the mixture was filtered through a glass-fiber paper (GF/A, Whatman), and an additional 50 mℓ of acetone was passed through the filter. The filtrate was placed into a 500 mℓ separatory funnel while the filter and its contents were homogenized with an additional 100 mℓ of acetone. Upon filtration, the residue was washed with 50 mℓ of acetone. To the acetone filtrates were added 100 mℓ of hexane and after mixing, 100 mℓ of water. After gentle swirling the hexane (upper phase) was transferred to a second funnel and the aqueous phase was extracted with 2 × 100 mℓ of hexane. After washing the pooled hexane extracts with 2 × 100 mℓ of water, the hexane was evaporated *in vacuo*.[117]

The E vitamers were eluted with diethyl ether-hexane (5:95) at a flow rate of 2 mℓ/min from a column of 5 μm LiChrosorb Si 60. The spectrofluorimeter was set at 290 nm excitation and 330 nm emission wavelengths. For certain samples (e.g., rubber latex), a saponification step was introduced and the nonsaponifiable fraction subjected to HPLC.[118] A similar method including the saponification step had been applied to resolve tocopherols from seed oils.[119]

To ensure recovery of α-tocopherol acetate, McMurray and Blanchflower[120] developed a protocol for plant tissues that included extraction and saponification preceding reverse-phase HPLC. Dry samples were milled before extraction, whereas wet samples such as silage were finely cut and mixed. The procedure is as follows. To 125 mℓ PTFE wide mouth bottles were added the samples (5 g), 10 mℓ of 5% ascorbic acid in 0.1 *N* HCl and 10 mℓ of ethanol. After mixing for 15 min on a reciprocating shaker, 40 mℓ of water and 40 mℓ of diethyl ether were added, and the bottles set on the shaker for an additional 1 hr. The phases were separated by centrifugation at 2000 g for 10 min and the upper ether layers were transferred to a 150 mℓ round bottom flask. The aqueous layer was reextracted with 40 mℓ of ether by shaking for 30 min. After centrifugation the ether layers were pooled and reduced to about 5 mℓ on a rotary evaporator. The ether solution was transferred to ground-glass stoppered 100 × 16 mm test tubes and evaporated to dryness with nitrogen at 50°C. The residue was treated with 3 mℓ of 2% ascorbic acid and brought to 70°C for a few minutes. After the addition of 7 mℓ of 60% aqueous KOH, the mixture was maintained at 70°C for 15 min. Upon cooling the mixture 4 mℓ of *n*-hexane and 3 mℓ of water were added and the tube was mixed vigorously for 1 min. Following centrifugation at 1000 g for 5 min to separate the phases, the hexane layer was separated and analyzed by HPLC.

Analyses were performed with a Waters HPLC instrument containing the Model 6000 pump, Model UK6 injector and a reverse-phase column (μBondapak C_{18}, 300 × 3.9 mm).

The mobile phase was methanol-water (95:5) at a flow rate of 3 mℓ/min and the eluates were detected with a fluorescence spectrophotometer and flow cell Perkin-Elmer Model 204 (Beaconsfield, England) using an excitation wavelength of 296 nm and an emission wavelength of 330 nm.[120]

Recoveries of α-tocopherol acetate ranged from 80 to 90%. Resolution of other isomers was not successful on this column with the indicated eluent. Fluorescence detection increased the selectivity of this method.[120]

Vatassery and Hagen have reported a method of isolation of α-tocopherol from mammalian tissue. Brain tissue (50 to 100 mg) was homogenized in a glass-teflon homogenizer in a solution of 1 mℓ of redistilled ethanol, by 2 mℓ of nanograde hexane and 1 mℓ of 10% aqueous ascorbic acid. A control was also prepared in which a standard solution of 1 μg of α-tocopherol replaced the tissue. The teflon pestle was operated by a mechanical drill at 1324 rpm for 2 min in the cold. After centrifugation of the homogenate at 1200 rpm for 10 min at 5°C to separate the phases, the upper hexane layer was transferred to anhydrous sodium sulfate to remove traces of water and analyzed by HPLC directly. The mixture was resolved on a 180 × 0.2 cm I.D. (1/8 in. O.D.) column of 37 to 50 μm Corasil II (Waters Associates) with a mobile phase of hexane-methanol (99.4:0.6) flowing at 0.5 mℓ/min.[121]

b. GC and TLC

Tissue analyses for the E vitamers employing TLC follows essentially the same protocols devised for biological fluids (see Section III.H.2.b). They represent adaptations of the original method by Bieri.[107] Vitamin E dimers in plant oils have been measured by a combination of TLC and reaction GC.[122] After saponification and TLC, the zones were scraped and injected onto a GC column. With the injector port maintained at 275°C, it served as a pyrolytic chamber converting the E dimers back to monomeric forms, which were separated on a column containing 3% SE-30.

Early GC studies on the E and K vitamers have been reviewed.[123-125]

c. Large Scale Sample Preparation

K Vitamers in liver are isoprenologs containing 4 to 13 isoprenyl residues. Dihydro, tetrahydro, and hexahydro isoprenyl derivatives were also found among the 12 to 19 menaquinones isolated depending upon the species. Matschiner and co-workers have developed protocols for the large-scale isolation and purification of these menaquinones from liver.[126] Although bioassay suggested only 10 to 16 μg/g tissue of phylloquinone was present, this method provided sufficient amounts to measure the mass spectra of the menaquinones. After the liver was ground and dehydrated with absolute ethanol, it was extracted with acetone. A series of columns were next used including silicic acid, eluted with benzene-hexane mixtures[127] and a reverse-phase system containing equal parts of hydrophobic celite and polyethylene powder, with hexane as the stationary phase and isopropyl alcohol, water and acetic acid mixtures as the mobile phase. Paraffin-coated and silver nitrate-impregnated silica gel plates were used in the last step (see Table TLC 33-35). Zones were then scraped from the plates, eluted from the silica gel with acetone-hexane (9:1) and subjected to mass spectrometry.

Menaquinone-4 was also isolated from the livers of chicks and rats using essentially the same method, except for the last step, in which GC was performed in addition to MS.[128] (For GC conditions see Table GC 32). GC was also used for the determination of menadione (unprenylated menaquinone, vitamin K_3) with OV-17 on Dexsil 300 as the stationary phase. This water-soluble vitamin is extracted with methanol which can then be injected directly.[129] The menadione bisulfate can be treated in the same manner and it undergoes pyrolysis on the column to regenerate menadione.[130]

Recently Seifert has isolated phylloquinone from plants by GC. Fresh leaves (1 to 2 kg) were lyophilized at plate temperatures of 27 to 32°C and an ultimate vacuum of 100 μm in a freeze dryer (Repp Industries, Inc., Gardiner, N.Y.). They were then mashed to fine particles and stored at −20°C protected from light. All extracts obtained in subsequent steps were maintained at room temperature and were protected from light.

A 30 g sample of dried leaves was subjected to Soxhlet extraction with 750 mℓ of hexane for 3 hr. A 1/3 aliquot of the hexane extract was reduced to 10 mℓ on a rotary evaporator and chromatographed on a 2.5 cm (I.D.) × 25 cm column containing 50 g of Woelm alumina (neutral activity, grade 1) with 3% by weight of water. This portion gave a suitable amount of vitamin K_1 (50 to 300 μg) for GC analyses. The hexane extract was eluted with 175 mℓ of each of the following: hexane, ether-hexane (1.5:98.5), ether-hexane (4:96), and ether-hexane (6:94 v/v). Most of the K_1 was detected in the first ether-hexane (1.5:98.5) eluate.

GC analyses were performed on a 2 mm I.D. × 2.1 m glass column packed with 80 to 100 mesh Chromosorb G (DMCS treated and acid-washed) coated with 2.5% by weight of Dexsil 300GC. The column extended into detector and injector ports providing an all glass system. The chromatograph was operated under the following conditions: column temperature, 290°C; injector temperature, 270°C; hydrogen flame detector temperature, 265°C; nitrogen carrier gas flow rate, 40 mℓ/min (at 25°C). K_1 eluted at about 26 min under these conditions and was identified by mass spectral analyses. Recoveries of added K_1 averaged 92 to 97%. Light sensitivity caused a 50% loss of K_1 in the plant samples after 9 days.[131]

Menaquinones have been isolated from bacteria using some of the above methods, but to remove contaminants such as isopropyl fatty acid esters, Campbell and Bentley[132] used acetone rather than isooctane-isopropyl alcohol as the extracting solvent.

Thus the wet paste (of *Mycobacterium phlei* cells) was homogenized in a Waring blender (ten bursts of 10 sec each) with three to four volumes of acetone. The solid material was filtered and recycled twice. The combined acetone extracts were reduced in volume to 10% and were then partitioned between ether and water, about ten volumes each. The yield of nonpolar lipid, obtained by separation, drying, and evaporation of the ether layer, was in the range 50 to 70 mg/100 g wet weight of paste.

It must be emphasized that the above operation, and all those that follow, must be conducted in virtual darkness if extensive *trans-cis* isomerism and decomposition of the menaquinones are to be avoided.

In a typical run, the nonpolar lipids (179.2 mg) were fractionated on a column of Sephadex LH-20 (130 g; 1.8 × 192 cm; flow rate, 34 mℓ/hr; swelling time, 22 hr) established in the solvent system isooctane-methanol-chloroform (2:1:1). After development of the column with 100 mℓ of the solvent, 3.5 mℓ fractions were collected. Ultraviolet spectrometry revealed that the characteristic chromophores of a 2,3-disubstituted naphthoquinone were restricted to fractions 3 to 14, with fractions 3 and 14 containing only trace amounts; yields: fractions 3 to 10 (MK_9 (II-H_2) and MK_{10} (II-H_2)), 91.1 mg; fractions 11 to 14 (MK_9 (II-H_2) and MK_8 (II-H_2)), 63.4 mg.

The combined fractions 11-14 (63.4 mg) from the above column were rechromographed on Sephadex LH-20 (130 g; 1.8 × 192 cm; flow rate, 14 mℓ/hr; swelling time, 24 hr) using the previously described solvent system. After development with 109 mℓ of solvent, 3.5 mℓ fractions were taken. Ultraviolet spectroscopy located menaquinone in fractions 1 to 5. Mass spectrometry indicated that fractions 4 and 5 (18.8 mg) were rich in MK_8 (II-H_2).

Final purification was achieved by repeated reversed-phase chromatography on layers of silica gel (250 μ) prepared with 0.02% Rhodamine 6G and impregnated with 1.5% paraffin oil. The developing solvent was acetone-water (19:1). The quinone was visualized by *very* brief exposure to ultraviolet light of wavelength 254 nm and was eluted from the silica gel with ethyl acetate. The paraffin oil contaminating the eluted quinone was removed by thin-layer chromatography on silica gel (25 μ) using cyclohexane-benzene (2:1) as solvent. Pure MK_8 (II-H_2) was obtained as an oil (about 1.5 mg).[132]

A similar preparative procedure was developed to isolate *E. coli* menaquinones.[133] A procedure using Permutit first developed by Doisy was adopted for the isolation of menaquinone-6 from anaerobic bacteria.[134]

A recent protocol for the isolation of bacterial menaquinones with tetrahydroisoprenoid side chains included 2 TLC steps. The procedure is as follows. Cells of *O. turbata* AJ 9191 were collected by centrifugation from 186 ℓ of culture media and lyophilized to give a dry weight of 303 g. The sample was treated with 3 ℓ of acetone-ether (4:1 v/v) and 500 mℓ

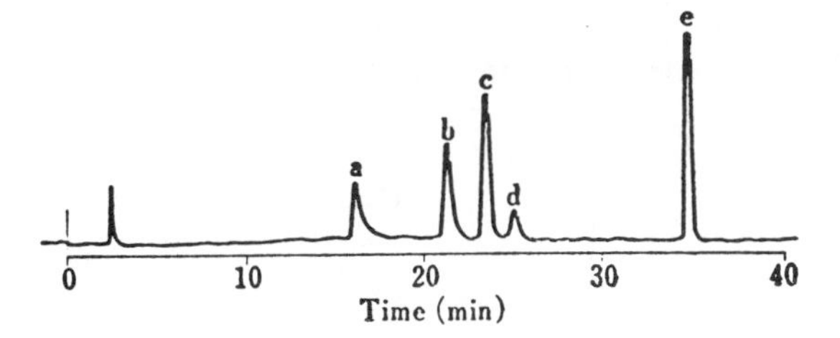

FIGURE 12. HPLC of chromanols.

Conditions: 6000 psi, ambient temperature, UV detection at 297 nm.

Column: Micro Pack Si (10 μm).

Solvent, flow rates: Gradient 5% of chloroform-ethyl acetate (8:2) in hexane increased 0.5% 1 min 40 mℓ/hr. a: 2,2,5,7,8-pentamethylchroman-6-ol, b: 2,2,5,7-tetramethylchroman-6-ol, c: 2,2,5,8-tetramethylchroman-6-ol, d: 2,2,7,8-tetramethyl-chroman-6-ol, e: 2,2-dimethylchroman-6-ol.[137]

aliquots were stirred for 3 hr. After repeating the extraction three times, the organic extracts were pooled, filtered, and evaporated to dryness *in vacuo*.

The residue was dissolved in minimal amounts of acetone for preparative TLC on silica gel plates (20 × 20 cm silica gel GF_{254}, Type 60, Merck, 0.5 mm thick). Authentic samples of vitamin K_1 and ubiquinone (from *Gluconobacter cerinus*) were placed on both sides of the plate. The plates were developed with hexane-benzene-chloroform (6:5:4) for 30 min. The yellow K_1 band appeared in the middle of the plates comigrating with the standard and resolved from carotenoids and ubiquinone, which appeared as a purple area on a yellow-green fluorescent background under UV light. The yellow band was scraped from the plate and washed with acetone to remove the menaquinone. The acetone solution was evaporated to dryness *in vacuo,* the residue taken up in a small aliquot of acetone, and the preparative TLC repeated.

The menaquinone containing mixture was then subjected to argentation preparative TLC using silica gel plates impregnated with 9% silver nitrate (w/w). Plates were developed with 2-butanone-hexane (1:9) and two yellow bands migrated to the middle of the plate. The upper band was identified as menaquinone-9 (H_6) by mass spectrometry. The lower band had spectral and chromatographic properties of menaquinone-9(H_4).

For *B. lipolyticum* AJ 1450, 270 g of lyophilized cells were obtained from 180 liters of culture media. When the above procedure was followed the argentation TLC afforded a single yellow band that was identified as menaquinone-8(H_4).[135]

Quinones have been isolated from bacteria and fungi simultaneously with polyprenols in protocols involving saponification in the presence of pyrogallol, alumina chromatography, and TLC (see Section III.E.2).

4. Synthetic Mixtures

a. Phosphorimetry

A phosphorescence study of K vitamers has been reported and the authors concluded that the method of analysis was superior to GC, TLC, polarographic, and colorimetric methods.[136]

HPLC has been used extensively to resolve isomeric mixtures of terpenoid quinones (see Table LC 18-20, Figure 12) as well as fat-soluble vitamers and their esters in pharmaceutical preparation.[138,139]

The official GC method of the Association of Analytical Chemistry for vitamin E has recently been modified to resolve acetate and succinate esters present in pharmaceuticals.[140]

Table 2
COMPOUNDS IDENTIFIED IN AN ESSENTIAL (YUZU) OIL

Compounds	% of Volatile components	Peak No. in GC	Methods of identification
α-Pinene	1.4		IR, GC
β-Pinene	0.5		IR, GC
Myrcene	2.2		IR, GC
Limonene	79.4		IR, GC
γ-Terpinene	9.5		IR, GC
Terpinolene	+		IR, GC
p-Cymene	0.2		IR, GC
p-Isopropenyl toluene	0.02	{1 ~ 10 {2 ~ 2	IR, GC, MS
β-Elemene	0.40	2 ~ 8	IR, GC, MS
γ-Elemene	0.51	2 ~ 11	IR, NMR, MS
δ-Elemene	0.30	2 ~ 4	IR, NMR, MS
α-Ylangene	+	2 ~ 6	IR, GC
β-Ylangene	+	2 ~ 7′	IR, GC
α-Copaene	0.30	2 ~ 7	IR, GC
β-Farnesene	0.86	{2 ~ 12 {1 ~ 30	IR, MS
β-Bisabolene	+	2 ~ 13	IR
α-Muurolene	+	2 ~ 15	IR, MS
γ-Cadinene	+	2 ~ 16	IR, MS
α-Curcumene	+	2 ~ 17	IR, MS
Calamenene	+	2 ~ 21	IR, MS
Caryophyllene	0.30	2 ~ 10	IR, GC, MS
Humulene	+	2 ~ 11	IR, GC, MS
Bicyclo-elemene	0.71	2 ~ 5	IR, MS, NMR
Thymol methyl ether	0.02	1 ~ 31	GC, MS
n-Hexyl aldehyde	+		GC
n-Octyl aldehyde	0.02	1 ~ 8	GC, MS
n-Nonyl aldehyde	+	1 ~ 9	GC, MS
n-Decyl aldehyde	0.04	1 ~ 15	GC, MS
n-Undecyl aldehyde	+	1 ~ 23	GC, MS
n-Dodecyl aldehyde	+	1 ~ 32	GC, MS
Cuminaldehyde	+	1 ~ 32	GC, MS
Perillaldehyde	0.03	1 ~ 33	GC, MS
Citral	0.01	1 ~ 27	IR, GC
Carvone	0.02	1 ~ 27	IR, GC
Methyl heptenone	+	1 ~ 10	GC
Citronellal	+	1 ~ 11	GC
Citronellyl acetate	+	1 ~ 23	GC, MS
Linallyl acetate	+	1 ~ 12	GC, MS
Citronellyl formate	+	1 ~ 22	GC, MS
Terpinyl acetate	+	1 ~ 19	GC, MS
Geranyl formate	+	1 ~ 28	GC, MS
Geranyl acetate	+	1 ~ 32	GC, MS
Perillyl acetate	+		GC, MS
p-Menthadien-1, 8(10)-ol-9 acetate	+		IR, MS
n-Nonyl alcohol	+		GC, MS
n-Amyl alcohol	+		GC
n-Hexyl alcohol	+		GC
Cumin alcohol	+	1 ~ 46	GC, MS
Citronellol	+	1 ~ 28	IR, GC

Table 2 (continued)
COMPOUNDS IDENTIFIED IN AN ESSENTIAL (YUZU) OIL

Compounds	% of Volatile components	Peak No. in GC	Methods of identification
Linalool	0.87	1 ~ 14	IR, GC
Geraniol	+	1 ~ 37	IR, GC
Nerol	+	1 ~ 36	GC
p-Menth-1,8-dien-4-ol	+	1 ~ 21	IR
α-Terpineol	0.12	1 ~ 25	IR, GC
Terpinen-4-ol	0.02	1 ~ 17	IR, GC
trans-Carveol	+	1 ~ 33	IR
cis-Carveol	+	{1 ~ 35 1 ~ 56	IR
Thymol	0.15	3-f	IR, MS
Nerolidol	+	3-c	IR
Elemol	+	3-e	IR, NMR
Globulol	+	3-g	IR, NMR
Spathulenol	+	3-h	IR, NMR
T-Cadinol	+	3-1	IR
T-Muurolol	+	3-m	IR
β-Eudesmol	+	3-o	IR, NMR
α-Cadinol	+	3-p	IR
Juniper camphor	+	3-s	IR
Capric acid	+		GC
Lauric acid	+		GC
Myristic acid	+		GC
Palmitic acid	+		GC
Stearic acid	+		GC
Aurapten			IR, NMR, MS
Phellopterin			IR, NMR, MS

REFERENCES

1. **Arpino, L., Vidal-Madjar, C., and Guiochon, G.,** *J. Chromatogr.*, 138, 173, 1977.
2. **DiCorcia, A., Liberti, A., Sambucini, C., and Samperi, R.,** *J. Chromatogr.*, 152, 63, 1978.
3. **Adams, R. P., Granat, M., Hogge, L. R., and von Rudloff, E.,** *J. Chromatogr. Sci.*, 17, 75, 1979.

3a. **Coscia, C. J.,** unpublished observations.

4. **Shinoda, N., Shiga, M., and Nishimura, K.,** *Agric. Biol. Chem.*, 34, 234, 1970.
5. **Jones, B. B., Clark, B. C., and Iacobucci, G. A.,** *J. Chromatogr.*, 178, 575, 1979.
6. **Kumamoto, J., Waines, J. G., Hollenberg, J. L., and Scora, R. W.,** *J. Agric. Food Chem.*, 27, 203, 1979.
7. **Croteau, R. and Karp, F.,** *Arch. Biochem. Biophys.*, 176, 734, 1976.
8. **Hood, L. V. S., Dames, M. E., and Barry, G. T.,** *Nature (London)*, 242, 402, 1973.
9. **von Rudloff, E.,** *Recent. Adv. Phytochem.*, 2, 127, 1969.
10. **Coscia, C. J., Botta, L., and Guarnaccia, R.,** *Arch. Biochem. Biophys.*, 136, 498, 1970.
11. **Guarnaccia, R., Botta, L., and Coscia, C. J.,** *J. Am. Chem. Soc.*, 96, 7079, 1974.
12. **Kodama, R., Noda, K., and Ide, H.,** *Xenobiotica*, 4, 85, 1974.
13. **Robertson, J. S. and Solomon, E.,** *Biochem. J.*, 121, 503, 1971.
14. **Kepner, R. E. and Maarse, H.,** *J. Chromatogr.*, 66, 229, 1972.
15. **Gueldner, T. C., Hutto, F. Y., Thompson, A. C., and Hedin, P. A.,** *Anal. Chem.*, 45, 376, 1973.
16. **Connell, D. W.,** *J. Chromatogr.*, 45, 129, 1969.
17. **Ikediobi, C. O., Hsu, I. C., Bamburg, J. R., and Strong, F. M.,** *Anal. Biochem.*, 43, 327, 1971.
18. **Maarse, H.,** *J. Chromatogr.*, 106, 369, 1975.
19. **Anderson, N. H., Falcone, M. S., and Syrdal, D. P.,** *Phytochemistry*, 9, 1341, 1970.
20. **Yamaguchi, I., Yokota, T., Yoshida, S., and Takahashi, N.,** *Phytochemistry*, 18, 1699, 1979.
21. **Powell, L. E. and Tautvydas, K.,** *Nature (London)*, 213, 292, 1967.
22. **Pitel, D. W., Vining, L. C., and Arsenault, G. P.,** *Can. J. Biochem.*, 49, 185, 1971.

23. **MacMillan, J. and Wels, C. M.,** *J. Chromatogr.,* 87, 271, 1973.
24. **Lindner, W. and Frei, R. W.,** *J. Chromatogr.,* 117, 81, 1976.
25. **Itoh, T., Tamura, T., and Matsumoto, T.,** *J. Am. Oil. Chem. Soc.,* 50, 300, 1973.
26. **Rozanski, A.,** *Analyst,* 97, 968, 1972.
27. **Higgins, J. W.,** *J. Chromatogr.,* 121, 329, 1976.
28. **Nachtmann, F., Spitzy, H., and Frei, R. W.,** *J. Chromatogr.,* 122, 293, 1976.
29. **Gfeller, J. C., Frey, G., and Frei, R. W.,** *J. Chromatogr.,* 142, 271, 1977.
30. **Erni, F. and Frei, R. W.,** *J. Chromatogr.,* 149, 561, 1978.
31. **Nigg, H. N., Thompson, M. J., Kaplanis, J. N., Svoboda, J. A., and Robbins, W. E.,** *Steroids,* 23, 507, 1974.
32. **Ikekawa, N., Hattori, F., Burbio-Lightbourn, J., Miyazake, H., Ishibashi, M., and Mori, C.,** *J. Chromatogr. Sci.,* 10, 233, 1972.
33. **Miyazaki, H., Ishibashi, M., Mori, C., and Ikekawa, N.,** *Anal. Chem.,* 45, 1164, 1973.
34. **Claude, J. R.,** *J. Chromatogr.,* 17, 596, 1965.
35. **Keenan, R. W., Rice, N., and Quock, R.,** *Biochem. J.,* 165, 405, 1977.
36. **Hemming, F. W.,** in *Biochemistry of Lipids,* Vol. 4, Goodwin, T. W., Ed., Butterworth, London, 1974.
37. **Keenan, R. W. and Kruczek, M.,** *Anal. Biochem.,* 69, 504, 1975.
38. **Wellburn, A. R., Stevenson, J., Hemming, F. W., and Morton, R. A.,** *Biochem. J.,* 102, 313, 1967.
39. **Burgos, J., Hemming, F. W., Pennock, J. R., and Morton, R. A.,** *Biochem. J.,* 88, 470, 1963.
40. **Dunphy, P. J., Kerr, J. D., Pennock, J. F., Whittle, K. J., and Feeney, J.,** *Biochim. Biophys. Acta,* 136, 136, 1967.
41. **Breckenridge, W. C., Wolfe, L. S., and Ng Ying Kin, N. M. K.,** *J. Neurochem.,* 21, 1311, 1973.
42. **Van Dessel, G., Lagrou, A., Hilderson, H. J., Dommisse, R., Esmans, E., and Dierick, W.,** *Biochim. Biophys. Acta,* 573, 296, 1979.
43. **Thorne, K. J. I. and Kodicek, E.,** *Biochim. Biophys. Acta,* 59, 280, 1966.
44. **Thorne, K. J. I. and Kodicek, E.,** *Biochem. J.,* 99, 123, 1966.
45. **Stone, K. J. and Strominger, J. L.,** *J. Biol. Chem.,* 247, 5107, 1972.
46. **Scher, M., Lennarz, W. J., and Sweeley, C. C.,** *Proc. Natl. Acad. Sci. U.S.A.,* 59, 1313, 1968.
47. **Wellburn, A. R. and Hemming, F. W.,** *J. Chromatogr.,* 23, 51, 1966.
48. **Stone, K. J., Wellburn, A. R., Hemming, F. W., and Pennock, J. F.,** *Biochem. J.,* 102, 325, 1967.
49. **Goldie, A. H. and Subden, R. E.,** *J. Chromatogr.,* 84, 192, 1973.
50. **Davies, B. H.,** in *Biochemistry of Lipids II,* Vol. 14, Goodwin, T. W., Ed., University Park Press, Baltimore, 1977.
51. **Stewart, L. and Wheaton, T. A.,** *J. Chromatogr.,* 55, 325, 1971.
52. **Davies, B. H.,** in *Chemistry and Biochemistry of Plant Pigments,* Vol. 1, 2nd ed., Goodwin, T. W., Ed., Academic Press, London, 1976.
53. **Moss, C. P. and Weedon, B. C. L.,** in *Chemistry and Biochemistry of Plant Pigments,* Vol. 1, 2nd ed., Goodwin, T. W., Ed., Academic Press, London, 1976.
54. **Liaaen-Jensen, S.,** in *Carotenoids,* Isler, O., Ed., Birkhäuser, Basel, 1971.
55. **Eskins, K., Schofield, C. R., and Dutton, H. J.,** *J. Chromatogr.,* 135, 217, 1977.
56. **Safta, M. and Ostrogovich, G.,** *J. Chromatogr.,* 69, 215, 1972.
57. **Kjosen, H., Arpin, N., and Liaaen-Jensen, S.,** *Acta Chem. Scand.,* 26, 3053, 1972.
58. **Kushwaha, S. C., Pugh, E. L., Kramer, J. K. G., and Kates, M.,** *Biochim. Biophys. Acta,* 260, 492, 1972.
59. **Puglisi, C. V. and De Silva, J. A. F.,** *J. Chromatogr.,* 120, 457, 1976.
60. **Anderson, D. G. and Porter, J. W.,** *Arch. Biochem. Biophys.,* 97, 509, 1962.
61. **Taylor, R. F. and Davies, B. H.,** *J. Chromatogr.,* 103, 327, 1975.
62. **Nybraaten, G. and Liaaen-Jensen, S.,** *Acta Chem. Scand. Ser. B,* 28, 584, 1974.
63. **Liaaen-Jensen, S.,** *Acta Chem. Scand.,* 19, 1166, 1965.
64. **DeLuca, H. F., Ed.,** *Handbook of Lipid Research,* Vol. 2, Plenum Press, New York, 1978.
65. **De Ruyter, M. G. M. and De Leenheer, A. P.,** *Clin. Chem. (Winston-Salem, N.C.),* 22, 1593, 1976.
66. **Puglisi, C. V. and DeSilva, J. A. F.,** *J. Chromatogr.,* 152, 421, 1978.
67. **DeRuyter, M. G. M. and De Leenheer, A. P.,** *Clin. Chem. (Winston-Salem, N.C.),* 24, 1920, 1978.
68. **Bieri, J. G., Tolliver, T. J., and Catignani, G. L.,** *Am. J. Clin. Nutr.,* 32, 2143, 1979.
69. **Abe, K., Ishibashi, K., and Ohmae, M.,** *Vitamins,* 51, 272, 1977.
70. **Hänni, R., Hervouet, D., and Busslinger, A.,** *J. Chromatogr.,* 162, 615, 1979.
71. **Frolich, C. A., Tavela, T., Peck, G. L., and Sporn, M. B.,** *Anal. Biochem.,* 86, 743, 1978.
72. **Frolich, C. A., Tavela, T., and Sporn, M. B.,** *J. Lipid Res.,* 19, 32, 1978.
73. **Garry, P. J., Pollack, J. D., and Owen, G. M.,** *Clin. Chem. (Winston-Salem, N.C.),* 16, 766, 1970.
74. **Pollack, D. J., Owen, G. M., Garry, P. J., and Clark, D.,** *Clin. Chem. (Winston-Salem, N.C.),* 19, 977, 1973.
75. **Bubb, F. A. and Murphy, G. M.,** *Clin. Chim. Acta,* 48, 329, 1973.

76. **Thompson, J. N., Erdody, P., Brien, R., and Murray, T. K.,** *Biochem. Med.*, 5, 67, 1971.
77. **Vahlquist, A. R.,** *Int. J. Vitam. Nutr. Res.*, 44, 375, 1974.
78. **Roberts, A. B., Nichols, M. D., Frolick, C. A., Newton, D. L., and Sporn, M. B.,** *Cancer Res.*, 38, 3327, 1978.
79. **Reitz, P., Wiss, O., and Weber, F.,** *Vitam. Horm. (N.Y.)*, 32, 327, 1974.
80. **Hänni, R., Bigler, F., Meister, W., and Englert, G.,** *Helv. Chim. Acta*, 59, 2221, 1976.
81. **Hänni, R. and Bigler, F.,** *Helv. Chim. Acta*, 60, 881, 1977.
82. **DeLucca, H. F., Zile, N. H., and Neville, P. F.,** *Lipid Chromatographic Analysis*, Vol. 2, Marcel Dekker, New York, 1969, 345-457.
83. **Ito, Y. L., Zile, M., Ahrens, H., and DeLuca, H. F.,** *J. Lipid Res.*, 15, 517, 1974.
84. **Paanakker, J. L. and Groenendijk, G. W. T.,** *J. Chromatogr.*, 168, 125, 1979.
85. **DeLuca, L. M.,** *Handbook of Lipid Research*, Vol. 2, DeLuca, H. F., Ed., Plenum Press, New York, 1978.
86. **Bhat, P. V., DeLuca, L. M., and Wind, M. L.,** *Anal. Biochem.*, 102, 243, 1980.
87. **Rodriguez, P., Bello, O., and Gaede, K.,** *FEBS Lett.*, 28, 133, 1972.
88. **McCormick, A. M., Napoli, J. L., Schnoes, H. K., and DeLuca, H.F.,** *Arch. Biochem. Biophys.*, 192, 577, 1979.
89. **Sietsma, W. K. and DeLuca, H. F.,** *Biochem. Biophys. Res. Commun.*, 90, 1091, 1979.
90. **Conrad, D. H. and Wirtz, G. H.,** *Immunochemistry*, 10, 273, 1973.
91. **Rotmans, J. P. and Kropf, A.,** *Vision Res.*, 15, 1301, 1975.
92. **Maeda, A., Shichida, Y., and Yoshizawa, T.,** *J. Biochem. (Japan)*, 83, 661, 1978.
93. **McKenzie, R. M., Hellwege, D. M., McGregor, M. L., Rockley, N. L., Riquetti, P. J., and Nelson, E. C.,** *J. Chromatogr.*, 155, 379, 1978.
94. **McCormick, A. M., Napoli, J. L., and DeLuca, H. F.,** *Anal. Biochem.*, 86, 25, 1978.
95. **Fenton, T. W., Vogtmann, H., and Clandinin, D. R.,** *J. Chromatogr.*, 77, 410, 1973.
96. **Bunnell, R. H.,** *Lipids*, 6, 245, 1971.
97. **De Leenheer, A. P., DeBevere, V. O., Cruyl, A. A., and Claeys, A. E.,** *Clin. Chem. (Winston-Salem, N.C.)*, 24, 585, 1978.
98. **Nilsson, B., Johansson, B., Jansson, L., and Holmberg, L.,** *J. Chromatogr.*, 145, 169, 1978.
99. **De Leenheer, A. P., DeBevere, V. O., DeRuyter, M. G. M., and Claeys, A. E.,** *J. Chromatogr.*, 162, 408, 1979.
100. **McMurray, C. H. and Blanchflower, W. J.,** *J. Chromatogr.*, 178, 525, 1979.
101. **Thompson, J. N., Erdody, P., and Maxwell, W. B.,** *Anal. Biochem.*, 50, 267, 1972.
102. **Bieri, J. G., Poukka, R. K. H., and Prival, E. L.,** *J. Lipid Res.*, 11, 118, 1970.
103. **Chiarotti, M. and Giusti, G. V.,** *J. Chromatogr.*, 147, 481, 1978.
104. **Lehman, J. and Slover, H. T.,** *Lipids*, 6, 35, 1971.
105. **Lovelady, H. G.,** *J. Chromatogr.*, 85, 81, 1973.
106. **Lin, S. N. and Horning, E. C.,** *J. Chromatogr.*, 112, 465, 1975.
107. **Bieri, J. G.,** in *Lipid Chromatographic Analysis*, Vol. 2, Marinetti, G. V., Ed., Marcel Dekker, New York, 1969.
108. **Hess, J. L., Pallansch, M. A., Harich, K., and Bunce, G. E.,** *Anal. Biochem.*, 83, 401, 1977.
109. **Lovelady, H. G.,** *J. Chromatogr.*, 78, 449, 1973.
110. **Bechtold, H. and Jähnchen, E.,** *J. Chromatogr.*, 164, 85, 1979.
111. **Duggan, D. E.,** *Arch. Biochem. Biophys.*, 84, 116, 1959.
112. **Hanse, L. G. and Warwick, W. J.,** *Am. J. Clin. Pathol.*, 46, 133, 1966.
113. **Thompson, J. N., Erdody, P., and Maxwell, W. B.,** *Biochem. Med.*, 8, 403, 1973.
114. **Kayden, H. J., Chow, C., and Bjornson, L. K.,** *J. Lipid Res.*, 14, 533, 1973.
115. **Okbayashi, H., Kanno, C., Yamauchi, K., and Tsugo, T.,** *Biochim. Biophys. Acta*, 380, 282, 1975.
116. **Shearer, M. J., McBurney, A., and Barkhan, P.,** *Vitam. Horm. (N.Y.)* 32, 513, 1974.
117. **Shearer, M. J., McBurney, A., Breckenridge, A. M., and Barkhan, P.,** *Clin. Sci. Mol. Med.*, 52, 621, 1977.
118. **Thompson, J. N. and Hatina, G.,** *J. Liq. Chromatogr.*, 2, 327, 1979.
119. **Abe, K., Yoguchi, Y., and Katsui, G.,** *J. Nutr. Sci. Vitaminol.*, 21, 183, 1975.
120. **McMurray, C. H. and Blanchflower, W. J.,** *J. Chromatogr.*, 176, 488, 1979.
121. **Vatassery, F. T. and Hagen, D. F.,** *Anal. Biochem.*, 79, 129, 1977.
122. **Gutfinger, T. and Letan, A.,** *Lipids*, 7, 483, 1972.
123. **Sheppard, A. J., Prosser, A. R., and Hubbard, W. D.,** *J. Am. Oil Chem. Soc.*, 49, 619, 1972.
124. **Sheppard, A. J.,** *Methods Enzymol.*, 18c, 461, 1971.
125. **Sheppard, A. J. and Hubbard, W. D.,** *Methods Enzymol.*, 18c, 465, 1971.
126. **Duello, T. J. and Matschiner, J. T.,** *Arch. Biochem. Biophys.*, 144, 330, 1971.
127. **Matschiner, J. T., Taggart, W. V., and Amelotti, J. M.,** *Biochemistry*, 6, 1243, 1967.

128. **Dialameh, G. H., Taggart, W. V., Matschiner, J. T., and Olson, R. E.,** *Internat. J. Vitam. Nutr. Res.*, 41, 391, 1971.
129. **Winkler, V. W.,** *J. Assoc. Off. Anal. Chem.*, 56, 1277, 1973.
130. **Winkler, V. W. and Yoder, J. M.,** *J. Assoc. Off. Anal. Chem.*, 55, 1219, 1972.
131. **Seifert, R. M.,** *J. Agric. Food. Chem.*, 27, 1301, 1979.
132. **Campbell, I. M. and Bentley, R.,** *Biochemisty,* 7, 3323, 1968.
133. **Campbell, I. M. and Bentley, R.,** *Biochemistry,* 8, 4651, 1969.
134. **Weber, M. M., Matschiner, J. T., and Peck, H. D.,** *Biochem. Biophys. Res. Commun.*, 38, 197, 1970.
135. **Yamada, Y., Inoue, G., Tahara, Y., and Kondo, K.,** *Biochim. Biophys. Acta,* 486 195, 1977.
136. **Aaron, J. J. and Winefordner, J. P.,** *Anal. Chem.*, 44, 2122, 1972.
137. **Matsuo, M. and Tahara, Y.,** *Chem. Pharm. Bull.*, 25, 3381, 1977.
138. **Williams, R. C., Schmit, J. A., and Henry, R. A.,** *J. Chromatogr. Sci.*, 10, 494, 1972.
139. **Dolan, J. W., Gant, J. R., Tanaka, N., Giese, R. W., and Karger, B. L.,** *J. Chromatogr. Sci.*, 16, 616, 1978.
140. **Sheppard, A. J. and Hubbard, W. D.,** *J. Pharm. Sci.*, 68, 98, 1979.

Index

INDEX

A

B

C

D

E

F

G

J

K

L

M

N

O

P

T

U

V

W

X

Y

Z